LABORATORY ATLAS

ANATOMY AND PHYSIOLOGY

Douglas J. Eder, PhD

Southern Illinois University—Edwardsville

Shari Lewis Kaminsky, DPM

John W. Bertram

with histology photos
by
Edward Reschke

 Mosby

St. Louis Baltimore Berlin Boston Carlsbad Chicago London Madrid
Naples New York Philadelphia Sydney Tokyo Toronto

Editor-in-Chief: James M. Smith
Editor: Robert J. Callanan
Project Manager: Carol Sullivan Wiseman
Senior Production Editor: Linda McKinley
Designer: Betty Schulz

Credits
Chapter 1: Edward Reschke
Chapter 2: David J. Mascaro & Associates
Chapter 3: John V. Hagen
Chapter 4: Douglas Eder, Shari Lewis Kaminsky, and John Bertram
Chapter 5: Figure 5-1 by Christine Oleksyk
    Tables from Thibodeau GA and Patton KT: *Anatomy and physiology*, ed 2, St Louis, 1993, Mosby.

Printed in the United States of America
Composition by Color Associates, Inc.
Printing/binding by Von Hoffman Press

Mosby–Year Book, Inc.
11830 Westline Industrial Drive
St. Louis, Missouri 63146

ISBN 0-8016-7051-9

94 95 96 97 98 / 9 8 7 6 5 4 3 2 1

CONTENTS

DEDICATION

To our spouses
Suzanne, Joe, and Mary Ellen

ACKNOWLEDGMENT

We obtained all animal material pictured from Nasco Company, Ft. Atkinson, Wisconsin. Cat cadavers were skinned at the factory and packed in a non-formaldehyde preservative. At our request, Nasco personnel selected particularly well-injected cadavers for us; we thank them for this service. We would also like to thank our colleague, Les Wiemerslage of Belleville Area Community College, who reviewed our animal dissections.

CHAPTER 1
HISTOLOGY

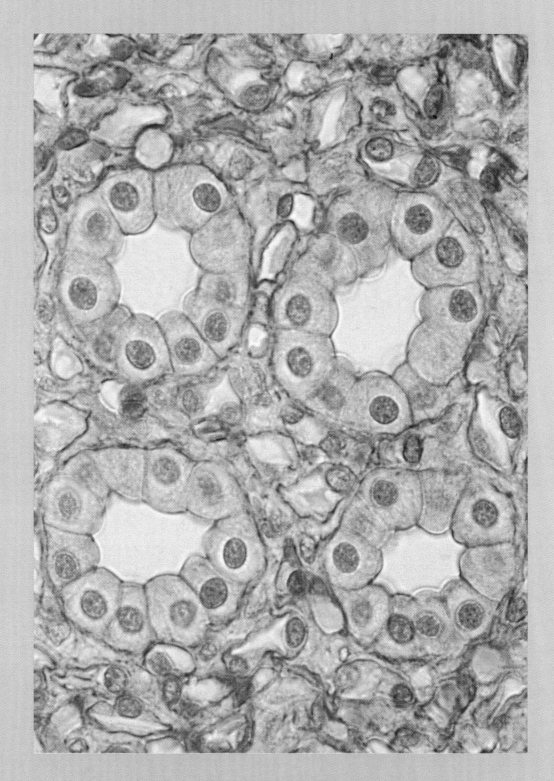

Simple Cuboidal Epithelium

Figure 1-1   Interphase
Nuclear membrane intact with chromatin visible. (×250.)

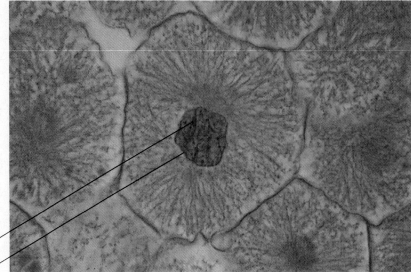

Chromatin

Nuclear membrane

Figure 1-2   Prophase
Duplicated chromosomes condensed into visible strands; nuclear membrane absent. (×250.)

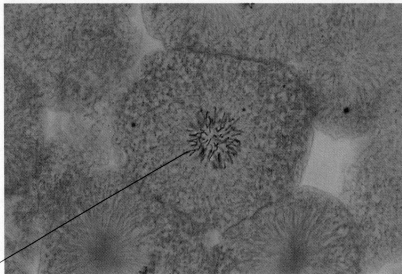

Chromosome

Figure 1-3   Metaphase
Darkly stained chromosomes positioned by microtubular framework to align at cell equator. Spindle fibers and aster visible. (×250.)

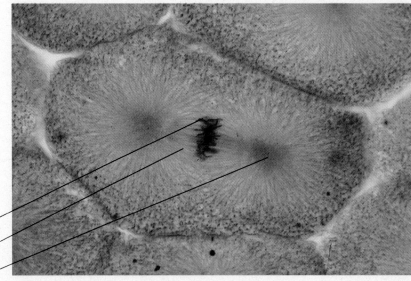

Chromosomes

Spindle fibers

Aster

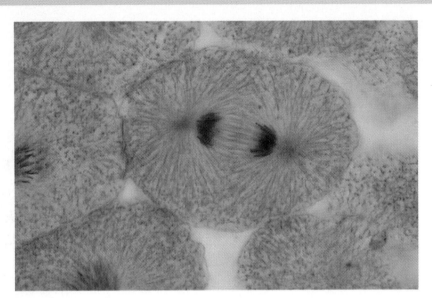

Figure 1-4   Anaphase
Darkly stained chromosomes move to opposite poles under microtubular influence. Spindle fibers and aster visible. (×250.)

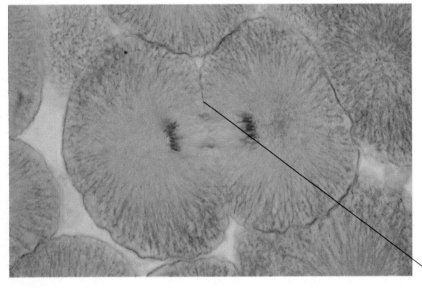

Figure 1-5   Telophase
Separated chromosomes lose microtubular attachments. Belt of actinomyosin forms at equator, assists in formation of new cell membranes and cytokinesis. Cleavage furrow forms two daughter cells. (×250.)

Cleavage furrow at equator

Figure 1-6 Simple Squamous Epithelium
Single layer of flat cells covering a surface.
From human omentum. (×250.)

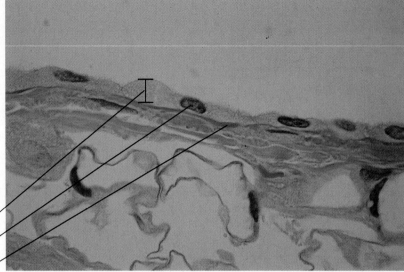

Simple squamous epithelium

Nucleus

Basement membrane

Figure 1-7 Simple Squamous Epithelium
Surface view of flattened cells. Human
mesothelium. (×250.)

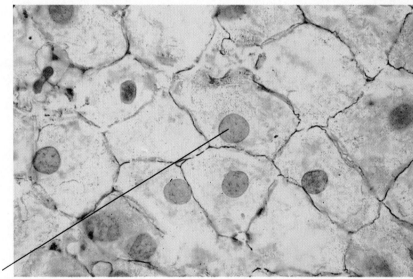

Nucleus

Figure 1-8 Simple Cuboidal Epithelium
Although not strictly cube shaped, cuboidal
cells are roughly equidimensional in length,
width, and depth. Single layer of cells lining
surface of kidney tubules. Cross section.
(×250.)

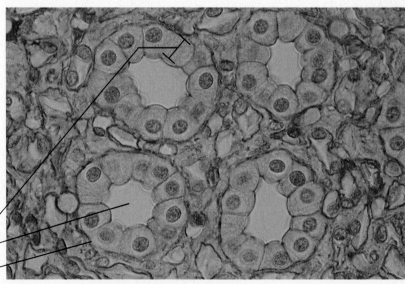

Simple cuboidal epithelium

Lumen of tubule

Basement membrane

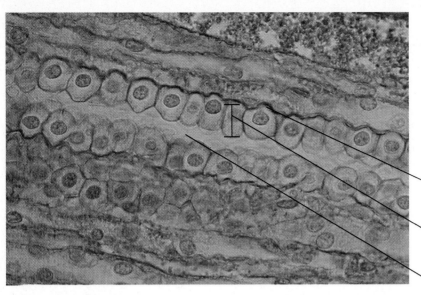

Figure 1-9 Simple Cuboidal Epithelium
Longitudinal section of kidney tubule.
(×250.)

Basement membrane

Simple cuboidal epithelium

Lumen of tubule

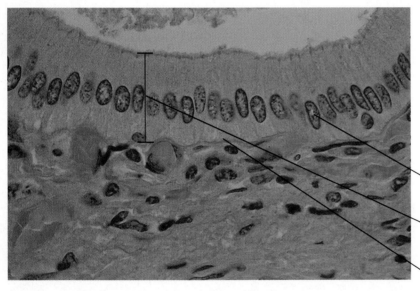

Figure 1-10 Simple Columnar Epithelium
Cellular height is much greater than width
or length. Nuclei generally appear in a row.
From pancreatic duct. (×250.)

Nucleus

Simple columnar epithelium

Basement membrane

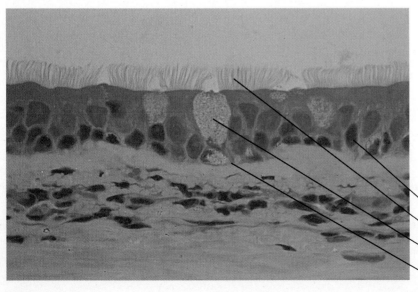

**Figure 1-11 Pseudostratified Ciliated
Columnar Epithelium**
Nuclei appear to lie in two rows, but in fact
all cells in single layer are in contact with
basement membrane. Section shows well-
defined cilia, three goblet cells, basement
membrane, underlying connective tissue.
From monkey trachea. (×100.)

Nucleus

Cilia

Goblet cell

Basement membrane

Figure 1-12 Pseudostratified Ciliated Columnar Epithelium

Section shows cilia, multiple layers of nuclei, basement membrane, underlying connective tissue. From human trachea. (×250.)

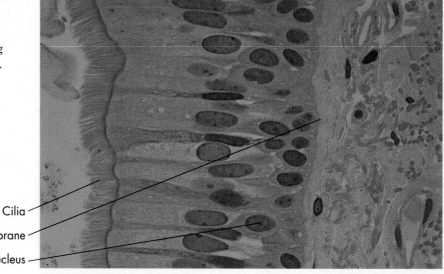

Cilia

Basement membrane

Nucleus

Figure 1-13 Stratified Squamous Epithelium

Flattened cells at surface change to less flattened morphology in deeper layers. Oral cavity of rabbit. (×100.)

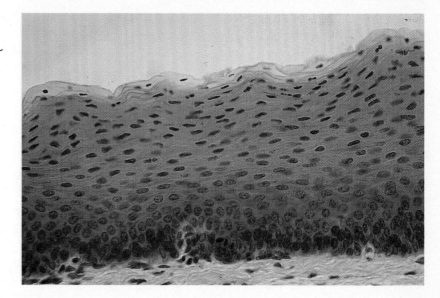

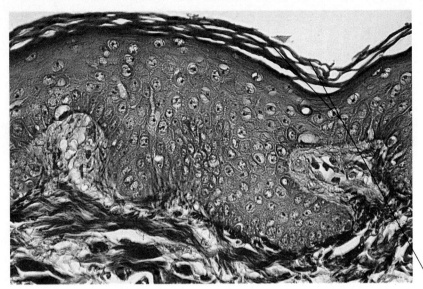

Figure 1-14   Stratified Squamous Epithelium

Flattened, keratinized cells at surface show variations in form in deeper layers. From papilla in human skin. (×100.)

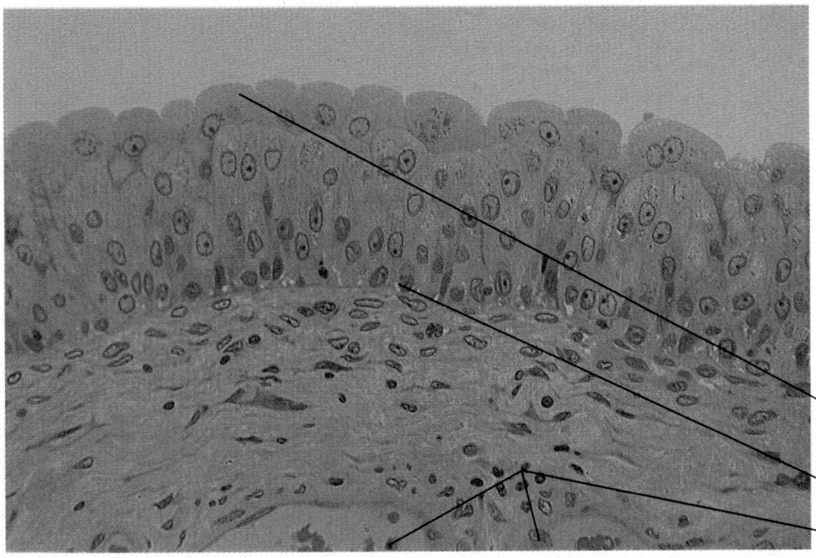

Keratinized cells

Figure 1-15 Transitional Epithelium from Urinary Bladder

Umbrella cells stretch and flatten as bladder fills. Basement membrane separates epithelium from underlying connective tissue containing blood vessels. (×250.)

Umbrella cell

Basement membrane

Blood vessels

Figure 1-16   Dense Regular Elastic Tissue
Extracellular elastin fibers running parallel in a plane. Structure permits tissue elasticity and recoil. From aorta. (×100.)

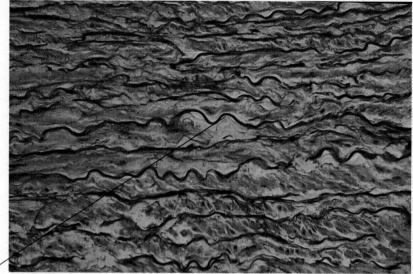

Elastin fiber

Figure 1-17   Reticular Connective Tissue
Mesh of collagen fibers appears as dark lines; provides scaffold for cellular organization of this lymph node. (×250.)

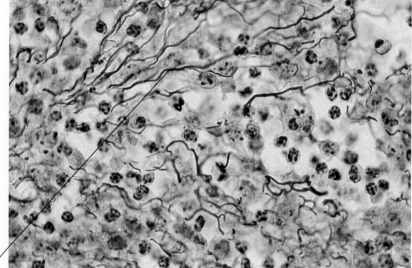

Collagen fiber

Figure 1-18   Loose (Areolar) Fibrous Connective Tissue
Thin, dark bands of collagen run in all directions through intercellular spaces of subcutaneous tissue, permit flexible resistance to mechanical stress. (×100.)

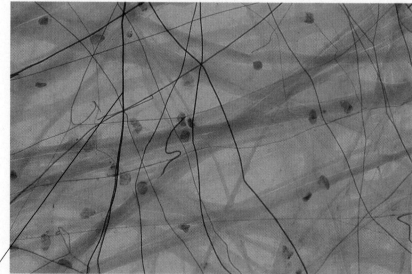

Collagen fiber

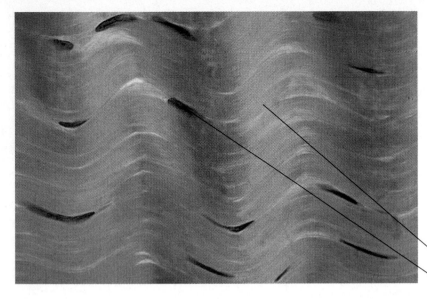

Figure 1-19 Dense Fibrous Connective Tissue

Thicker bands of collagen running in regular, parallel rows resist mechanical stress mainly along course of fibers. Monkey tendon. (×250.)

Collagen fibers

Nucleus of fibroblast

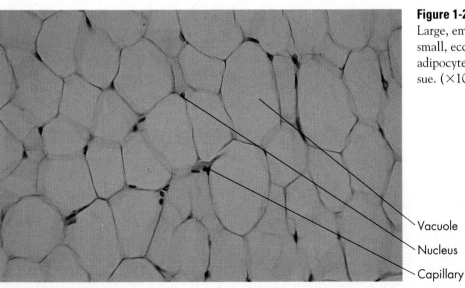

Figure 1-20 Dense Fibrous Connective Tissue

Thicker bands of collagen running in irregular rows give multidirectional tensile strength. Collagen-secreting fibroblasts appear throughout. From aponeurosis. (×100.)

Nuclei of fibroblasts

Collagen fibers

Figure 1-21 Adipose Tissue

Large, empty, polyhedral vacuoles dominate small, eccentrically located cell nuclei of adipocytes. Fine capillaries run through tissue. (×100.)

Vacuole

Nucleus

Capillary

Figure 1-22   Fibrocartilage
Cell nests of chondrocytes in territorial matrix surrounded by coarse extracellular fibers. (×250.)

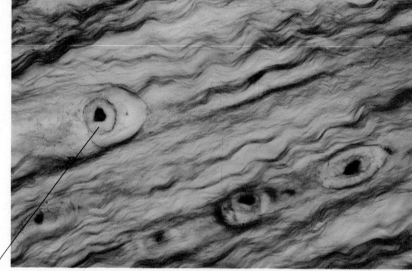

Chondrocyte

Figure 1-23   Hyaline Cartilage
During growth, cells often form small clusters and move apart as they secrete extracellular matrix. (×100.)

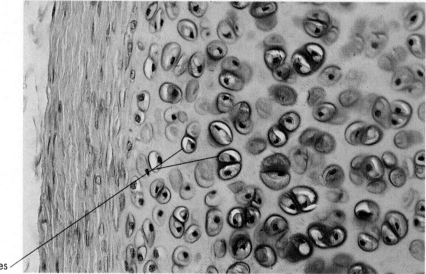

Chondrocytes

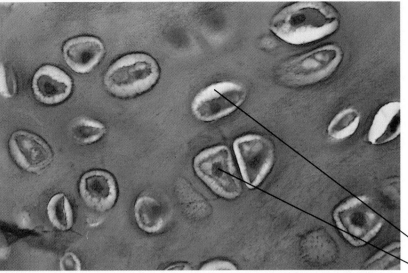

Figure 1-24 Hyaline Cartilage
Artifactual vacuolation forms characteristic lacuna around cell bodies. From trachea. (×250.)

Lacuna
Chondrocyte

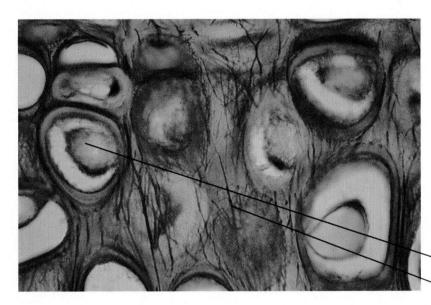

Figure 1-25 Elastic Cartilage
Extracellular matrix contains elastic fibers that confer elastic recoil to this tissue. (×250.)

Chondrocyte
Elastin fiber

Figure 1-26 Skin
Thick, keratinized, multilayered **stratum corneum** rests atop grainy **stratum granulosum** (stratum lucidum not clearly evident). **Stratum spinosum,** composed of irregularly shaped cells with indistinct nuclei, lies atop single, clearly nucleated **stratum basale**. Human palm. (×100.)

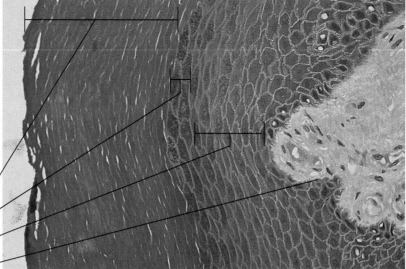

Stratum corneum

Stratum granulosum

Stratum spinosum

Stratum basale

Figure 1-27 Skin
Squamous epidermis with cornified layers overlying darkly stained stratum basale and connective tissue of underlying dermis. Single papilla visible. Human scalp. (×100.)

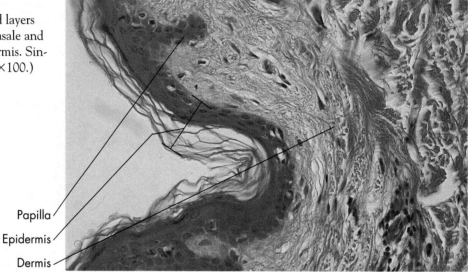

Papilla

Epidermis

Dermis

Figure 1-28 Meissner's Corpuscle in Dermis
Elongated oval body located in dermis just below stratum basale is thought to be responsible for part of fine touch reception. (×100.)

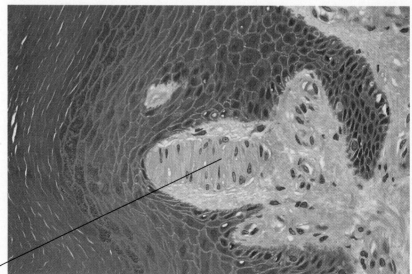

Meissner's corpuscle

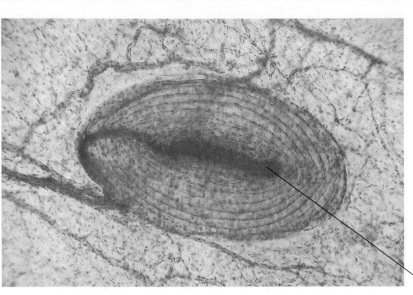

Figure 1-29   Pacinian Corpuscle
Encapsulated nerve ending found deep in dermis and throughout interior of body detects pressure. (×25.)

Free nerve ending

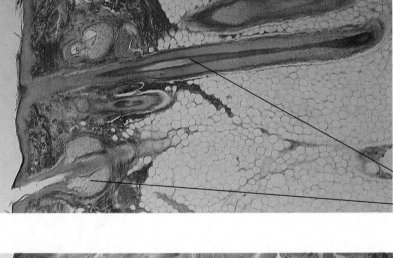

Figure 1-30   Human Scalp with Hair Follicle
Follicle root, with sheath embedded in pale adipose tissue, has sebaceous glands surrounding it near surface. (×10.)

Hair follicle

Sebaceous gland

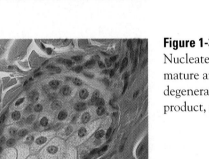

Figure 1-31   Detail of Sebaceous Gland
Nucleated germinative cells at base of gland mature and accumulate lipid. At duct, they degenerate and lyse to release their oily product, sebum. (×100.)

Sebaceous gland

Hair follicle

Figure 1-32   Compact Bone
Center of "tree ring" structure, Haversian canal, contains blood vessel. Osteocytes imprisoned in small, dark lacunae surrounding central Haversian canal receive nutrition and communicate via canaliculi, or little canals. Human. (×50.)

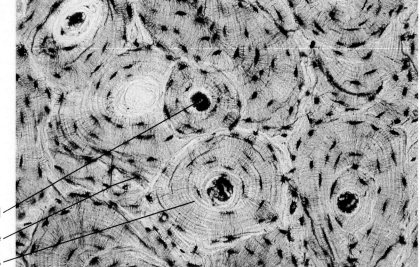

Haversian canal
Lacunae
Canaliculus

Figure 1-33   Detail of Compact Bone
Haversian system evident.

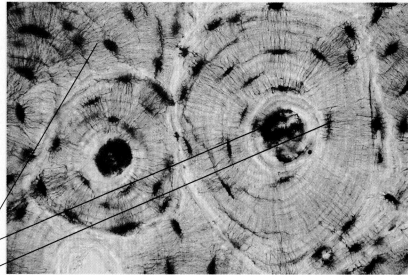

Canaliculus
Haversian canal
Osteocyte in lacuna

Figure 1-34   Cancellous (Spongy) Bone
Osteoblasts in spongy bone are engaged in secretion of new bony matrix. (×100.)

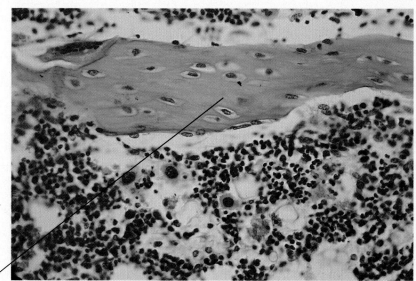

Spongy bone

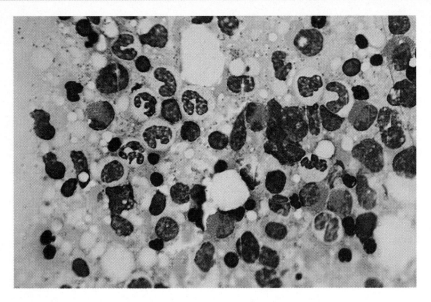

Figure 1-35   Red Bone Marrow
Medullary cavity of long bones contains stem cells and precursors to red and white cells and platelets. Human. (×250.)

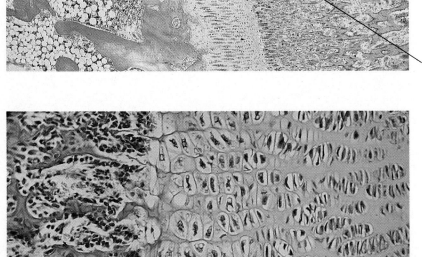

Figure 1-36   Developing Bone at Epiphyseal Line
Middle belt of cartilage undergoing primary calcification transforms into new bone. Secondary ossification occurs later.

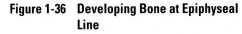

Cartilage

Figure 1-37   Detail of Epiphyseal Line
Epiphyseal plate cartilage at right transforms into zones of proliferating chondrocytes with primary ossification. Newly formed bone appears at left. (×50.)

Figure 1-38 Striated (Skeletal) Muscle (Cross Section)

Eccentrically located multiple nuclei accompany individual cells (fibers), each of which contains many myofibrils. Human tongue. (×250.)

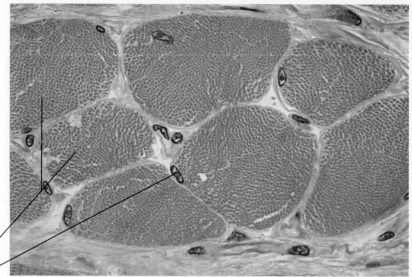

Muscle fibers

Nucleus

Figure 1-39 Striated (Skeletal) Muscle Fiber (Longitudinal Section)

Banded appearance arises from regular arrangement of overlapping bundles of thick and thin filaments (myosin and actin, respectively). Eccentrically located nuclei are thin and elongated. (×250.)

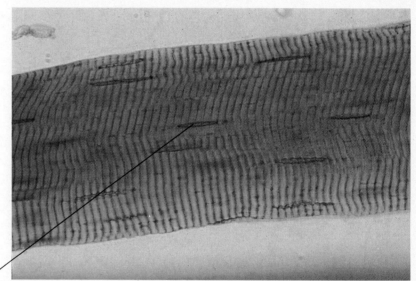

Nucleus

Figure 1-40 Striated (Skeletal) Muscle Fibers (Longitudinal Section)

Each light (I) band has a dark (Z) line through it. Each dark (A) band has a light (H) zone through it. (×250.)

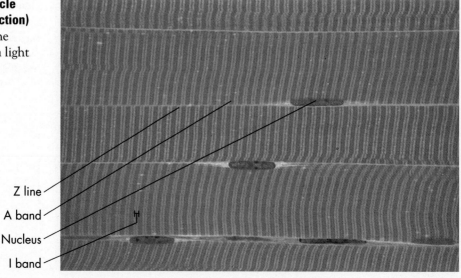

Z line

A band

Nucleus

I band

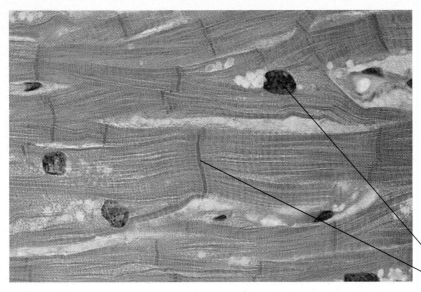

Figure 1-41   Cardiac Muscle (Longitudinal Section)

Multinucleated, striated muscle fibers branch and anastomose at junctions marked by dark intercalated disks. (×250.)

Nucleus

Intercalated disk

Figure 1-42   Smooth Muscle (Longitudinal Section)

Canoe- or spindle-shaped muscle cells lack striations, and each has a single, elongated nucleus. (×250.)

Nucleus

Figure 1-43   Innervation of Skeletal Muscle: Motor Endplate

Branching nerve bundle terminates in small, specialized dents, the **myoneural junctions**. Nerve terminals release small quantities of chemical neurotransmitter to stimulate muscle contraction.

Myoneural junction

Figure 1-44   Astrocytes (Neuroglia)

Star-shaped supporting cells of central nervous system modulate ionic environment. Cytoplasmic extensions make contact with blood vessel. Cat. (Silver stain; ×280.)

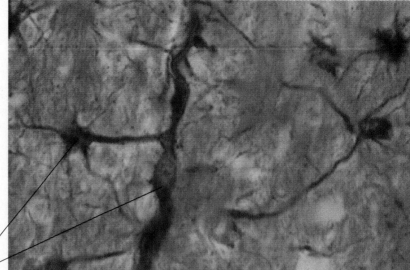

Astrocyte

Blood vessel

Figure 1-45   Purkinje Cells (Neurons)

Numerous branched processes (dendrites) receive information for processing. Single process (axon) sends information to other neurons. Human cerebellum. (×100.)

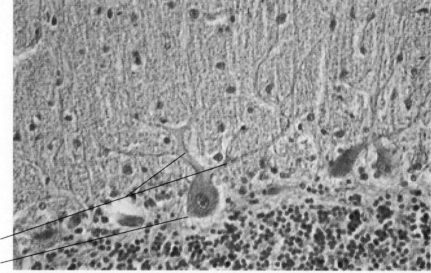

Dendrites

Axon

Figure 1-46   Pyramidal Cells

Neurons from human cortex directly receive information from hundreds of other cells; send information on to hundreds of others. (Fox-Golgi stain; ×100.)

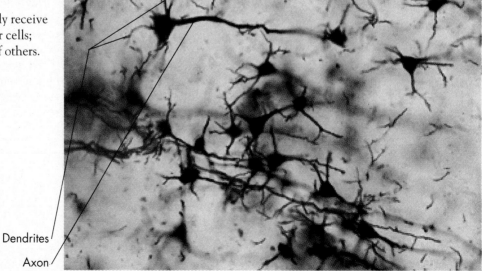

Dendrites

Axon

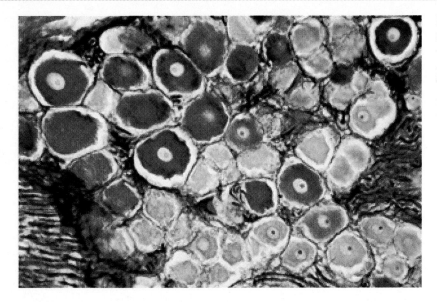

Figure 1-47   Dorsal Root Ganglion
Sensory signals representing pain, tempera-ture, pressure, muscle tension, joint position, and others depend on these cells. Their dendrites and axons collect sensory information throughout the body and route it into the spinal cord. (×100.)

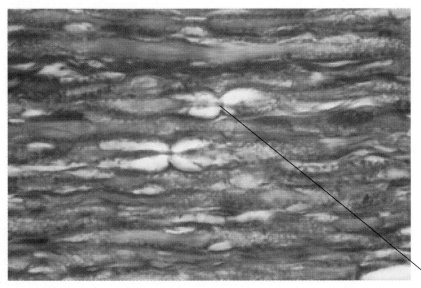

Node of Ranvier

Figure 1-48   Nerve Fibers (Longitudinal Section)
Clear areas show dimpling characteristic of nodes of Ranvier. (×250.)

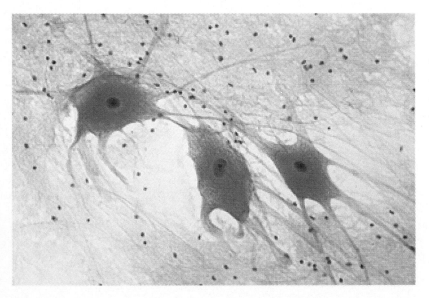

Figure 1-49   Motor Neurons of the Spinal Cord
Integrated command information from the brain and sensory signals enter these cells, whose efferent activity controls muscular contraction. Numerous synapses occur on dendrites and soma. (×50.)

Figure 1-50   Myelinated Nerve Fibers (Cross Section)

Central core stains dark; insulating myelin appears white. (×250.)

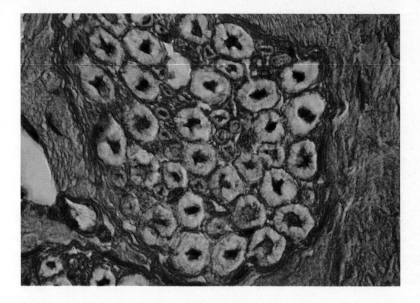

Figure 1-51   Spinal Cord, Lumbar Region (Cross Section)

Top is dorsal, bottom is ventral. Light central dot is central canal. Darkly staining H-shaped region is grey matter of cell bodies; surrounding lighter material is composed of myelinated axons. Ventral horns of grey matter contain motor neurons; dorsal horns contain cell bodies of sensory pathways. (×4.)

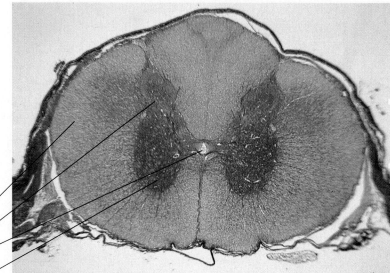

White matter
Dorsal horn
Central canal
Ventral horn

Figure 1-52   Retina

Layered structure evident. Dark line of cells near top is pigment epithelium. Broad striped region represents photoreceptors (rods and cones), whose nuclei stain heavily immediately beneath. Below receptor nuclei lie synaptic region and a layer of nuclei belonging to bipolar cells. Bipolar cell output synapses onto ganglion cells, only a few of which appear near bottom. Axons of ganglion cells form optic nerve. (×100.)

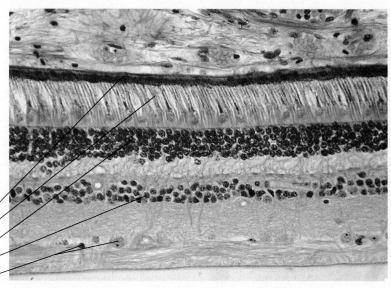

Pigmented epithelium
Rods and cones
Receptor nuclei
Bipolar cell nuclei
Ganglion cells

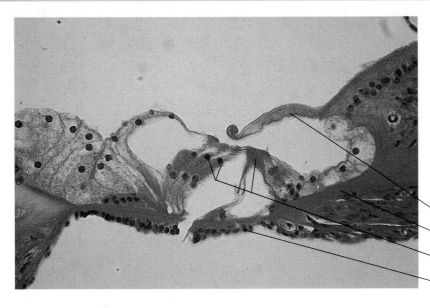

Figure 1-53 Organ of Corti
Thick finger of tectorial membrane extends from right to stimulate complex of four hair cells (three on left, one on right) of central structure that rests on important basilar membrane. Nerve fibers from hair cells exit right to spiral ganglion for processing and transmission of messages to brain. (×100.)

— Tectorial membrane

— Nerve fibers

— Hair cells

— Basilar membrane

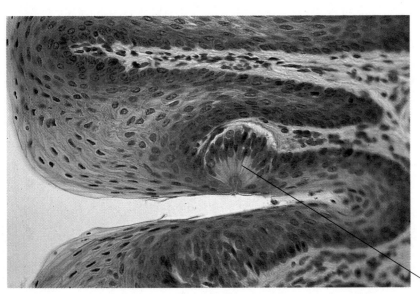

Figure 1-54 Taste Bud
Dissolved chemicals enter fungiform papilla through small pore to directly stimulate sensory cells and initiate taste perception. (×100.)

— Taste bud

Figure 1-55   Thyroid Gland Follicles

Cuboidal epithelium surrounds endocrine follicles of the thyroid gland, the only gland that stores substantial amounts of its own hormone. (×100.)

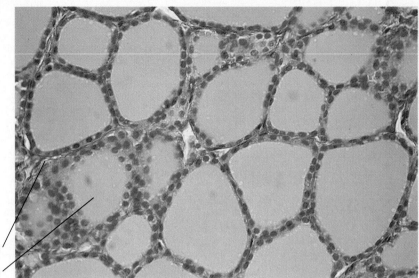

Cuboidal cells

Thyroid follicle

Figure 1-56   Pancreas

The pancreatic islet cells form the endocrine portion of the pancreas. Alpha cells secrete glucagon, beta cells secrete insulin, and delta cells secrete somatostatin. The exocrine portion of the pancreas secretes digestive enzymes through a series of ducts.

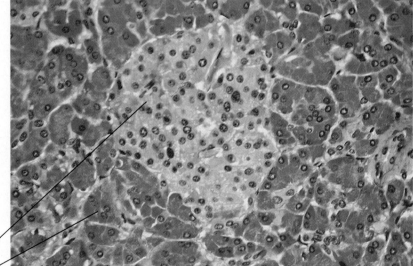

Pancreatic islet

Exocrine portion of pancreas

Figure 1-57   Adrenal Cortex

Outer zone of rounded groups of cells (zona glomerulosa) secretes mineralcorticosteroids (aldosterone). Middle area of cell stripes (zona fasciculata) secretes glucocorticosteroids. Innermost zone of cells arranged in a meshwork (zona reticulata) secretes mainly androgens. (×50.)

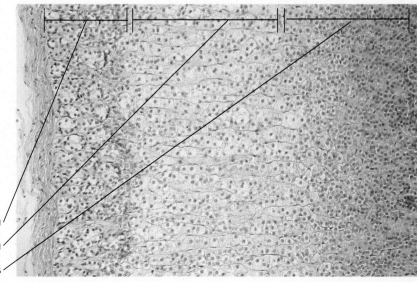

Zona glomerulosa

Zona fasciculata

Zona reticularis

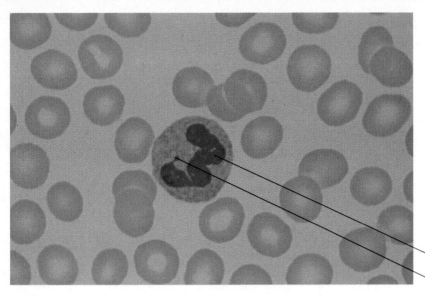

Figure 1-58   Neutrophil
Most numerous (65%) of the leukocytes, it is characterized by a multilobed nucleus and granular cytoplasm. Engages in phagocytosis. (Neutral dyes stain; ×640.)

Nucleus
Barr body

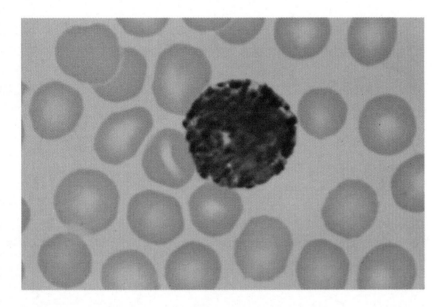

Figure 1-59   Basophil
Normally the rarest (1%) of the leukocytes, its kidney-shaped nucleus may be almost obscured by cytoplasmic granules. These cells contain numerous chemicals involved in inflammation. (Basic dyes stain; ×640.)

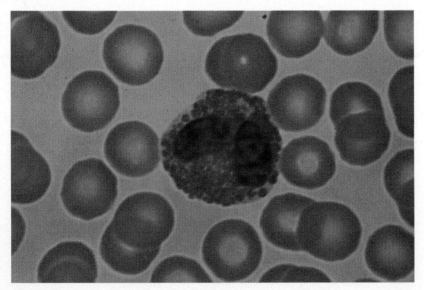

Figure 1-60   Eosinophil
Relatively rare (6%) leukocyte. Usually identifiable because of red-to-orange-staining cytoplasmic granules. Function not definitely known but elevated especially in allergies. (Selective eosin stain; ×640.)

Figure 1-61   Lymphocyte
Common (25%). Characterized by single-lobed, "dented" nucleus surrounded by clear cytoplasm. May be large or small. Heavily involved in the immune response including, for those resident in lymph nodes, synthesis of antibodies. (×640.)

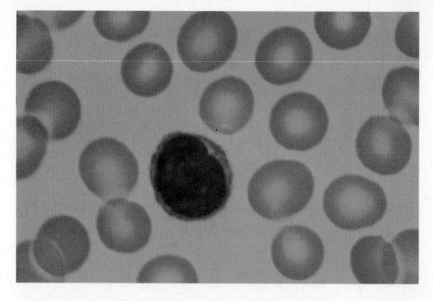

Figure 1-62   Monocyte
Relatively rare (3%). Lobed, often kidney-shaped nucleus is surrounded by clear cytoplasm. Largest of the leukocytes, this cell is a scavenger and engages in phagocytosis. (×640.)

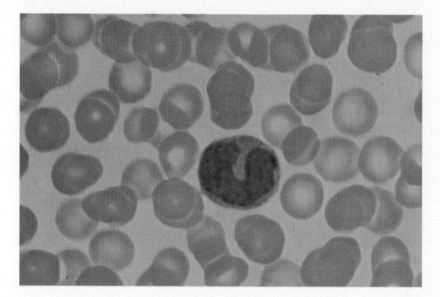

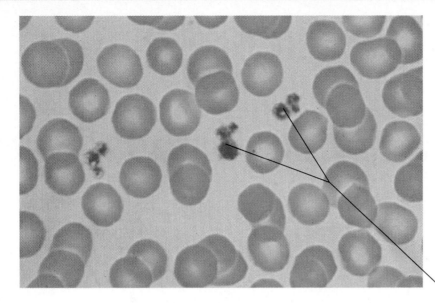

Figure 1-63 Erythrocytes (Red Blood Cells) and Platelets

Circulating erythrocytes are far more common than any of the leukocytes. Normally they have no nucleus but do contain the red pigment hemoglobin, which permits them to transport oxygen and carbon dioxide throughout the body. Typically they assume the shape of a biconcave disk. Their diameter of about 7 microns is useful for comparing sizes of other histological structures. Platelets are cellular remnants of a leukocyte precursor. These remnants contain numerous chemicals, including those important for clotting and inflammation. Platelets initiate blood clotting by forming a plug at wound sites. (×500.)

Platelets

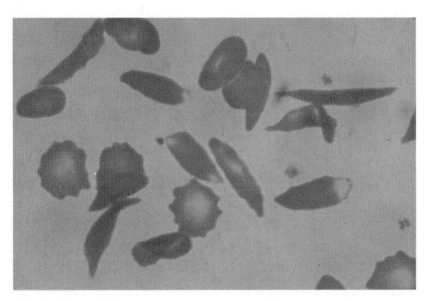

Figure 1-64 Sickle Cell Anemia

Genetic alteration of hemoglobin results in altered membrane structure and abnormal wavy or elongated, curved shape that often resembles a sickle *(upper left)*. Oxygen-carrying capacity is much reduced. (×500.)

Figure 1-65 Artery *(A)* and Vein *(V)*
Blood vessels possess a **tunica intima** that lines the lumen, outside of which is a muscular **tunica media**, and a connective tissue covering, the **tunica adventitia**. The tunica media of arteries is typically much thicker than that of veins. (×100.)

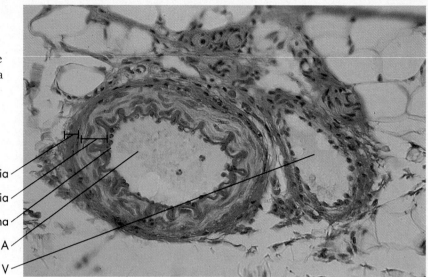

Tunica adventitia
Tunica media
Tunica intima
A
V

Figure 1-66 Arterial Cross Section
Single layer of darkly stained cells, the tunica intima, lines the lumen. Thick tunica media is composed of canoe-shaped smooth muscle cells. Outer adventitial layer of connective tissue provides elastic support and strength. (×50.)

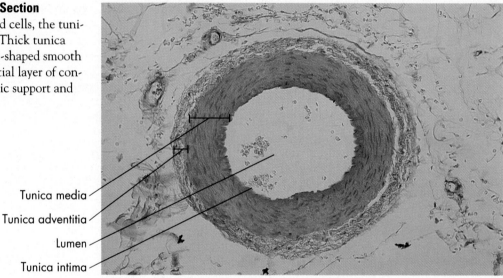

Tunica media
Tunica adventitia
Lumen
Tunica intima

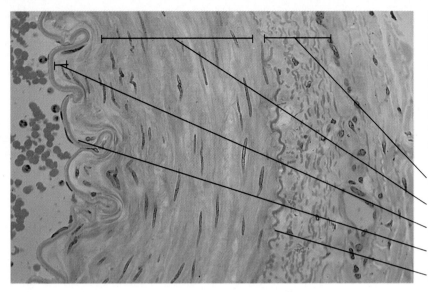

Figure 1-67 Detail of Arterial Wall
Inner endothelial cells of tunica intima (*left*) lie on a basement membrane. A thin layer of smooth muscle cells and elastic tissue (lamina propria) throws this tunic into folds. The tunica media contains multiple layers of smooth muscle cells regularly arranged. A wavy external elastic membrane separates the tunica media from the adventitia.

Adventitia

Tunica media

Lamina propria

Endothelium and basement membrane

External elastic membrane

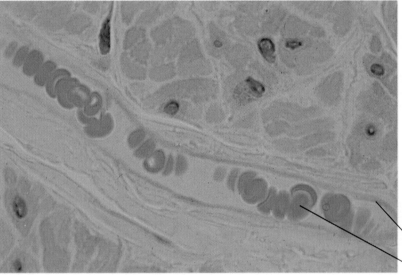

Figure 1-68 Capillary with Red Blood Cells in Single File
Capillary wall is made of flattened endothelial cells without complex tunics, a simple structure that facilitates the exchange of gases, nutrients, wastes, and hormones. (×400.)

Endothelium

Red blood cell

Figure 1-69   Lymph Node
Outer cortex containing several follicles surrounds medulla, with its narrow, dark medullary cords. Notch is hilum, through which blood and lymphatic vessels pass. (×5.)

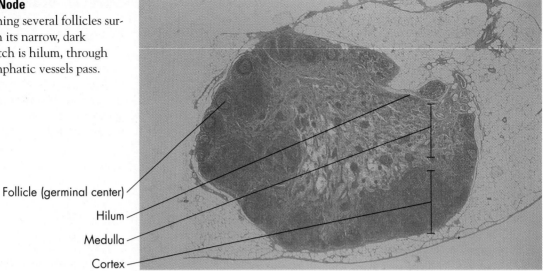

Follicle (germinal center)

Hilum

Medulla

Cortex

Figure 1-70   Valve of Lymphatic Vessel
One-way flow of lymph, from left to right in this figure, is ensured by valve action in lymph vessel. Vessels themselves are thin walled and lack musculature; pumping action occurs through compression by neighboring muscles. (×25.)

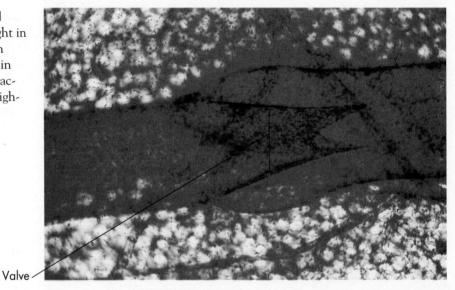

Valve

Figure 1-71   Thymus
Various lobules contain thick, darkly staining cortex surrounding a smaller, lighter-staining medulla. Small, round cellular patches in medulla are Hassall's corpuscles. In adults, much of thymus degenerates and is replaced by adipose tissue. (×10.)

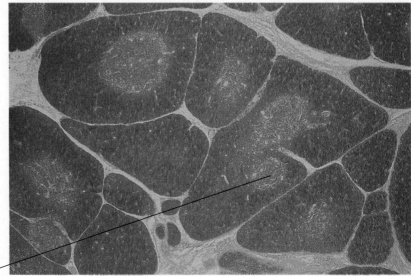

Hassall's corpuscle

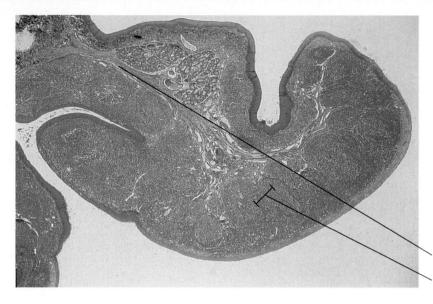

Figure 1-72   Palatine Tonsil
Outer capsule surrounds subcapsular sinus, under which are several large, rounded germinal centers (lymph nodes) surrounding travecular arteries and veins. Efferent lymph vessel leads out to upper left. (×5.)

Lymph vessel
Germinal center

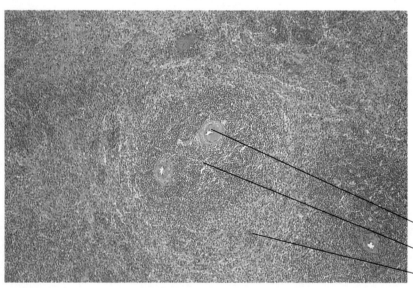

Figure 1-73   Spleen
Central blood vessels are surrounded by area of densely staining white pulp composed of lymphoid cells. Less densely staining red pulp, with fewer cell nuclei, surrounds white pulp. (×25.)

Blood vessels
White pulp
Red pulp

Figure 1-74   Alveoli
Thin-walled respiratory exchange surfaces
aid in rapid diffusion of gases. Bronchiole
terminates at atrium, which acts as entryway
into several individual alveolar sacs, greatly
multiplying surface area. (×50.)

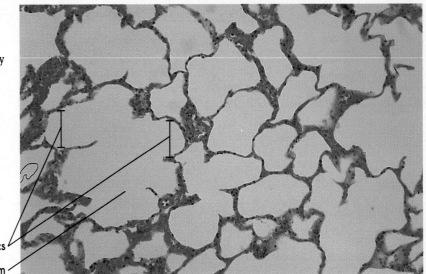

Alveolar sacs

Atrium

Figure 1-75   Detail of Alveolus
Squamous cells compose alveolar sac, which
is penetrated by thin-walled blood vessels
(*upper left*) containing erythrocytes. (×100.)

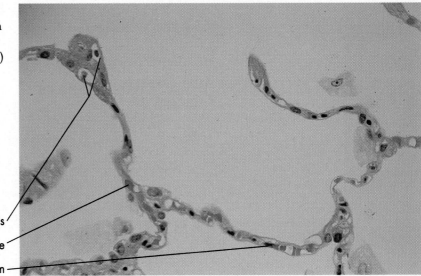

Blood vessels

Erythrocyte

Simple squamous epithelium

Figure 1-76   Bronchiole
Epithelial layer that lines the lumen is surrounded by layer of smooth muscle, which
regulates bronchiolar diameter. Round
structures outside of smooth muscle layer are
blood vessels. (×100.)

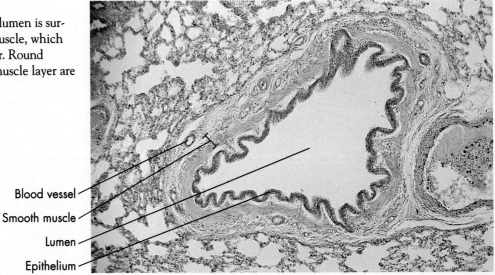

Blood vessel

Smooth muscle

Lumen

Epithelium

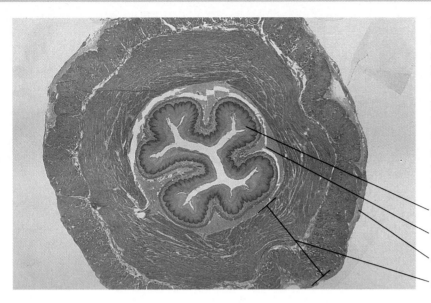

Figure 1-77   Esophagus

Surrounding the lumen, esophageal structure contains, in order, the four basic layers of the alimentary canal: **mucosa** (composed of epithelium, the thick lamina propria, and dark muscularis), **submucosa** (light with spaces, blood vessels, and lymph channels), two thick layers of the **muscularis** (circular and longitudinal), and the thin, connective **adventitia** on the surface. Cross section, human. (×3.)

Mucosa

Submucosa

Adventitia

Muscularis

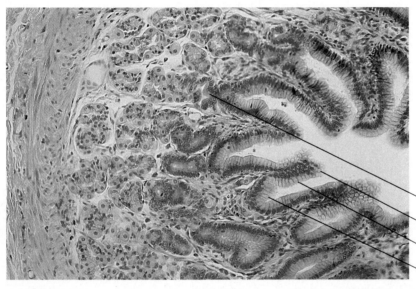

Figure 1-78   Stomach Mucosa

Visible at entrances to gastric pits are mucus-secreting goblet cells of columnar epithelium. Deeper in pits are acid-secreting parietal cells and enzyme-secreting chief cells. Endocrine-secreting cells near tip of pits are noncolumnar and smaller, with dark, round nuclei. Gastric pits penetrate deep into submucosal layer. Edge of muscularis layer is visible. (×50.)

Endocrine cells

Goblet cells

Parietal cells

Chief cells

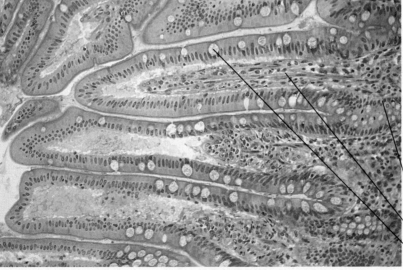

Figure 1-79   Small Intestine, Villi of Ileum
(Longitudinal Section)

Numerous pale goblet cells punctuate columnar epithelium that composes each villus. Core of villus contains small blood vessels and blind lymph channel (lacteal). Deep in crypts are endocrine cells, identifiable as dark, round nuclei in a noncolumnar cytoplasm. Human. (×50.)

Endocrine cells

Blood vessel and lacteal

Goblet cell

Figure 1-80   Small Intestine, Villi of Ileum (Cross Section)

Goblet cells emptying contents through brush border surface are evident. Core of villus contains blood vessels, lymph channels, and lymphocytes. Human. (×100.)

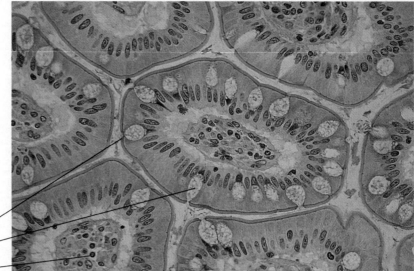

Brush border

Goblet cell

Lymphocyte

Figure 1-81   Large Intestine (Colon) (Cross Section)

Surface is thrown into folds but devoid of villi. Thick submucosa contains blood vessels and lymph channels. (×10.)

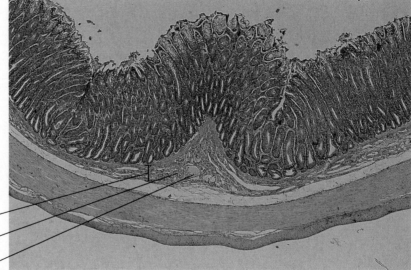

Submucosa

Blood vessel

Lymph channel

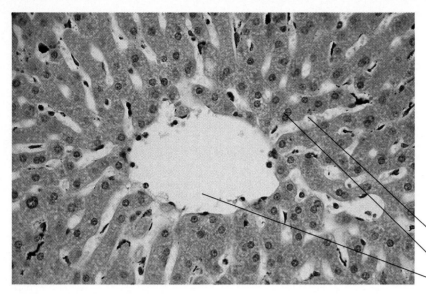

Figure 1-82   Liver with Central Vein and Sinusoids
Parenchymal hepatocytes lie in radial arrangement around central vein that is lined with single endothelial layer. Cords of hepatocytes are separated by spaces (sinusoids). Sinusoidal surface is covered by microvilli. (×100.)

Sinusoid

Hepatocyte

Central vein

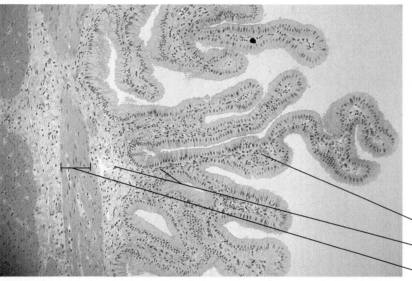

Figure 1-83   Gallbladder
Mucosal folds are covered by epithelium with well-developed microvilli. Lamina propria contains blood vessels. (×25.)

Blood vessel

Lamina propria

Muscularis

Figure 1-84 Vermiform Appendix (Cross Section)
Overall structure resembles that of colon. Large, darkly staining structures are lymphoid follicles, the size and number of which decrease with age. Human. (×3.)

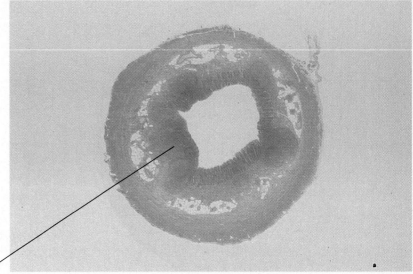

Lymphoid follicle

Figure 1-85 Sublingual Salivary Gland
Large, pale, mucus-secreting cells, some with caps of serous demilunes, secrete their contents into ducts that may be lined with striated epithelial cells indicative of ion exchange activity. (×100.)

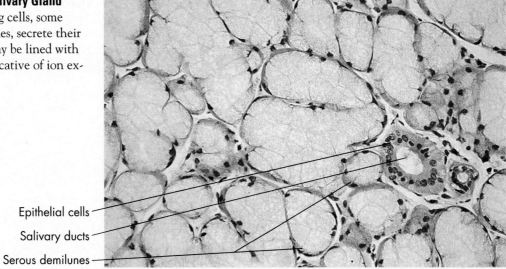

Epithelial cells

Salivary ducts

Serous demilunes

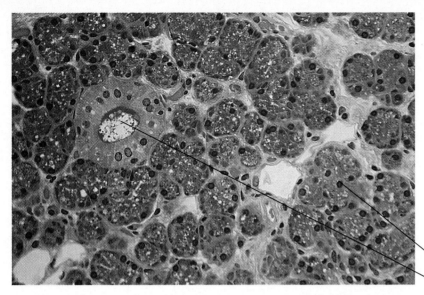

Figure 1-86 Parotid Salivary Gland
Granular serous cells with numerous, large, zymogen granules surround duct. Several tiny ducts run between clusters within the plane of section. Human. (×100.)

Zymogen granule

Salivary duct

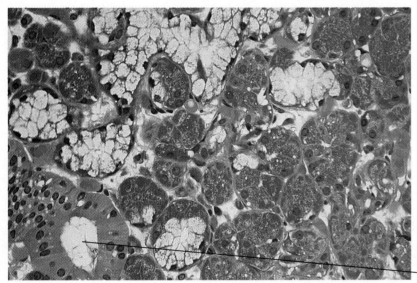

Figure 1-87 Submandibular Salivary Gland with Mucous (Light Staining) and Serous (Dark Staining) Components
Striated duct is visible at lower left. (×100.)

Duct

Figure 1-88   Bowman's Capsule and Glomerulus

Tuft of capillaries, surrounded by podocytes, protrudes into space of Bowman's capsule. Parietal surface is lined with single layer of simple squamous cells. (×100.)

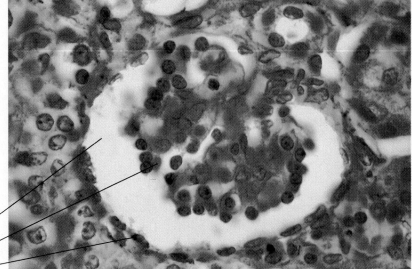

Space in Bowman's capsule

Glomerulus

Squamous cell

Figure 1-89   Two Glomeruli and Bowman's Capsules

"Lacy" edges of glomerulus on right shows characteristics of pregnancy-induced hypertension (PIH), here induced experimentally in a pregnant rat. (×50.)

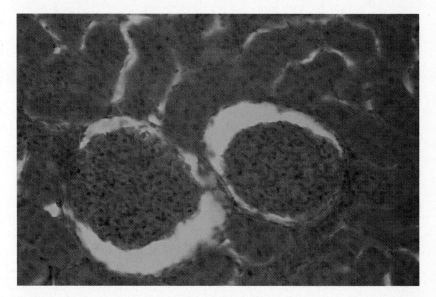

Figure 1-90   Distal Convoluted Tubules Lined with Cuboidal Epithelium

Cross section of rat. (×400.)

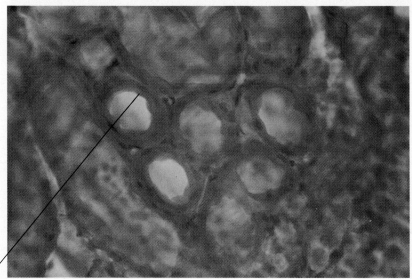

Cuboidal cell

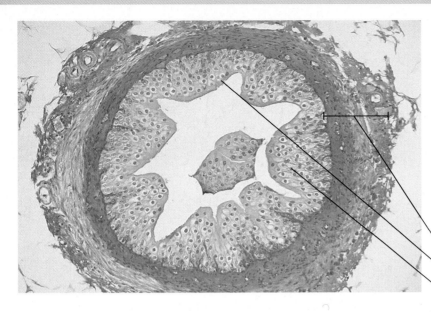

Figure 1-91 Ureter
Star-shaped lumen is lined with transitional epithelium that varies in thickness to change shape as lumen stretches. Delicate lamina propria separates epithelium from alternating layers of circular and longitudinal smooth muscle. (×25.)

Smooth muscle
Transitional epithelium
Lamina propria

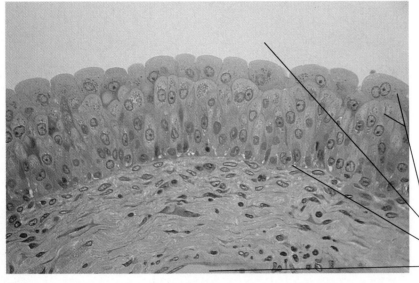

Figure 1-92 Bladder
Umbrella cells of transitional epithelium stretch and flatten as bladder fills. Basement membrane separates epithelium from underlying connective tissue containing blood vessels. Monkey. (×100.)

Umbrella cells
Free surface of bladder
Basement membrane
Blood vessel

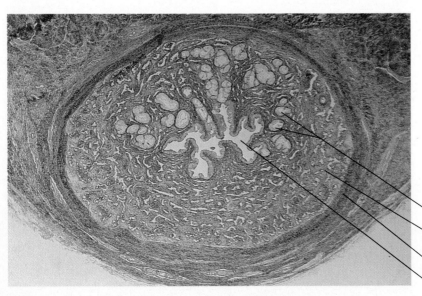

Figure 1-93 Urethra (within Penis)
Lumen is lined with transitional epithelium and is embedded in corpus spongiosum of the penis. Paraurethral glands located above the lumen in the figure secrete mucus into the urethra. A smooth muscle layer (tunica muscularis) surrounds the urethral structures. (×10.)

Paraurethral glands
Corpus spongiosum
Tunica muscularis
Lumen

Figure 1-94   Seminiferous Tubules of Testis Lined with Sertoli Cells and Germinativum in Various Stages of Development

Tunica propria surrounds each tubule. Interstitial spaces contain blood vessels and clumps of interstitial (Leidig) cells that secrete testosterone. (×50.)

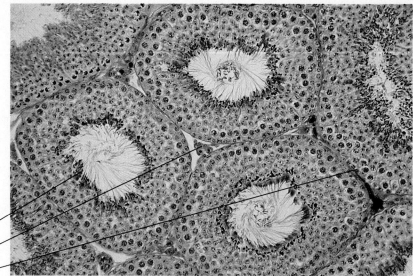

Sertoli cells

Interstitial cells

Tunica propria

Figure 1-95   Spermatozoa

Head contains numerous enzymes and nucleus with DNA. Thick midpiece just behind head is packed with mitochondria. (×250.)

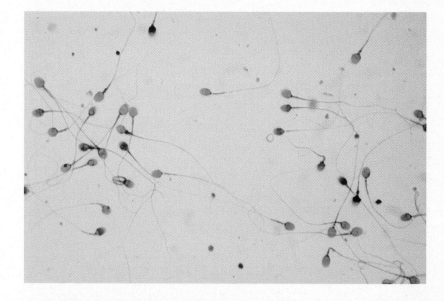

Figure 1-96   Epididymis

Tall, pseudostratified columnar epithelium with microvilli surrounds a lumen packed with clumps of spermatozoa. Narrow band of smooth muscle cells encircles each tubule.

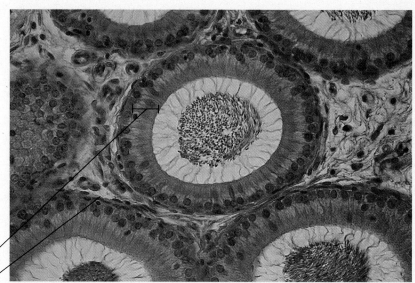

Pseudostratified columnar epithelium

Smooth muscle

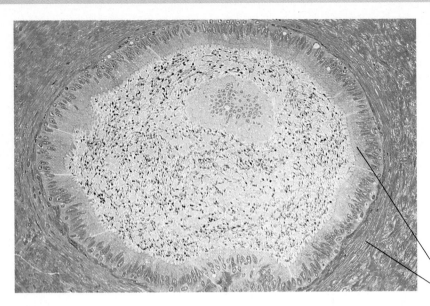

Figure 1-97 Ductus Deferens
Ciliated columnar epithelial cells line a spermatozoa-filled lumen. Three layers of smooth muscle cells surround mucosa, a circular layer between two longitudinal ones. (×50.)

Columnar epithelium of mucosa

Smooth muscle

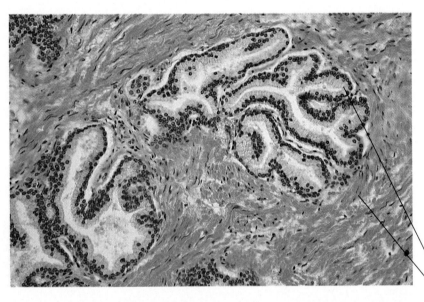

Figure 1-98 Prostate Gland
Mucosal surfaces, lined with tall columnar cells and darkly stained basal nuclei, are arranged in numerous deep folds. Lumina open directly into prostatic urethra. Smooth muscle and fibrocollagenous stroma surround luminal structures. Human. (×50.)

Columnar epithelium

Smooth muscle and fibrocartilage

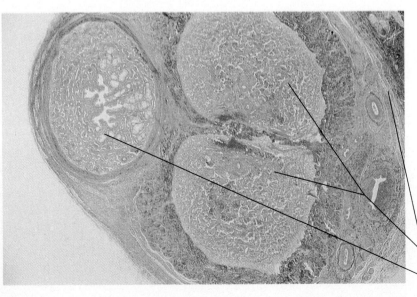

Figure 1-99 Penis
Two corpora cavernosa lie superior to single corpus spongiosum containing penile urethra. Septum between corpora cavernosa is incomplete. Dense fibrous connective tissue, tunica albuginea, surrounds the three vascular cavernosa. The inferior aspect appears on the left, the superior aspect on the right. (×5.)

Tunica albuginea

Corpora cavernosa

Corpus spongiosum

Figure 1-100   Ovary with Numerous Primordial Follicles and Two Primary Follicles

Primordial follicles contain oocytes that are not stimulated to complete the first meiotic division. Two primary follicles each contain ovum with nucleus and clear surrounding cytoplasm. Thin, clear **zona pellucida** is surrounded by a ring of even cuboidal cells, the **corona radiata**. (×25.)

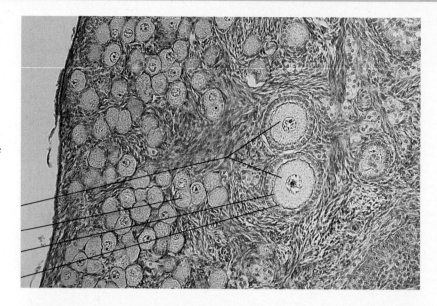

Figure 1-101   Detail of Oocyte in Primordial Follicle

Clear nucleus contains well-defined nucleolus. Neither zona pellucida nor corona radiata is evident. (×250.)

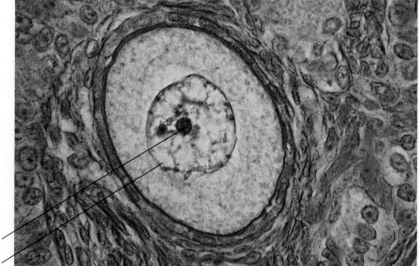

Nucleolus

Nucleus

Figure 1-102   Secondary Ovarian Follicle with Ovum

Bright zona pellucida surrounds outer membrane of ovum and in turn is surrounded by dark, cellular corona radiata. A large **antrum** has formed where the egg is not anchored to the follicular wall of **granulosa cells**. (×100.)

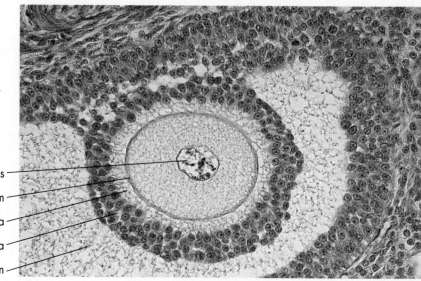

Nucleus

Membrane of ovum

Zona pellucida

Corona radiata

Antrum

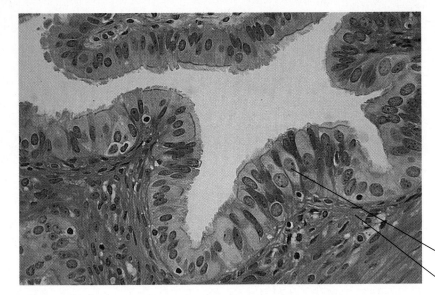

Figure 1-103   Fallopian (Uterine) Tube
Extensive folding of mucosa, lined with ciliated columnar epithelium, is common. Epithelium rests on thin basement membrane and flat connective tissue layer. Rhythmic beating of cilia helps transport ovum toward uterus; cell structure also suggests secretory function. Human. (×100.)

Columnar epithelium

Connective tissue

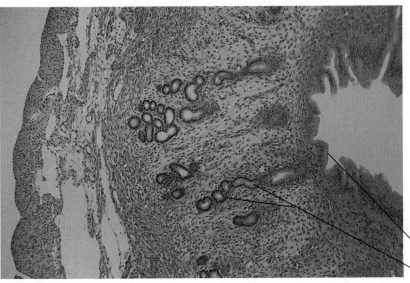

Figure 1-104   Uterus
Endometrial lining (*right*) during proliferative phase of uterine cycle shows thickening of epithelial surfaces and numerous coiled glandular ducts. (×25.)

Endothelial lining

Glandular ducts

CHAPTER 2
HUMAN SKELETAL ANATOMY

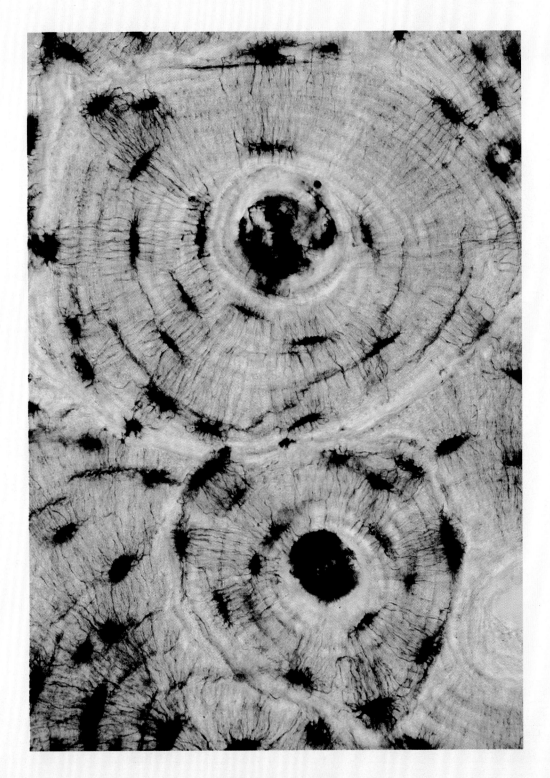

Detail of Compact Bone

A

B

Axial skeleton

Appendicular skeleton

Axial skeleton

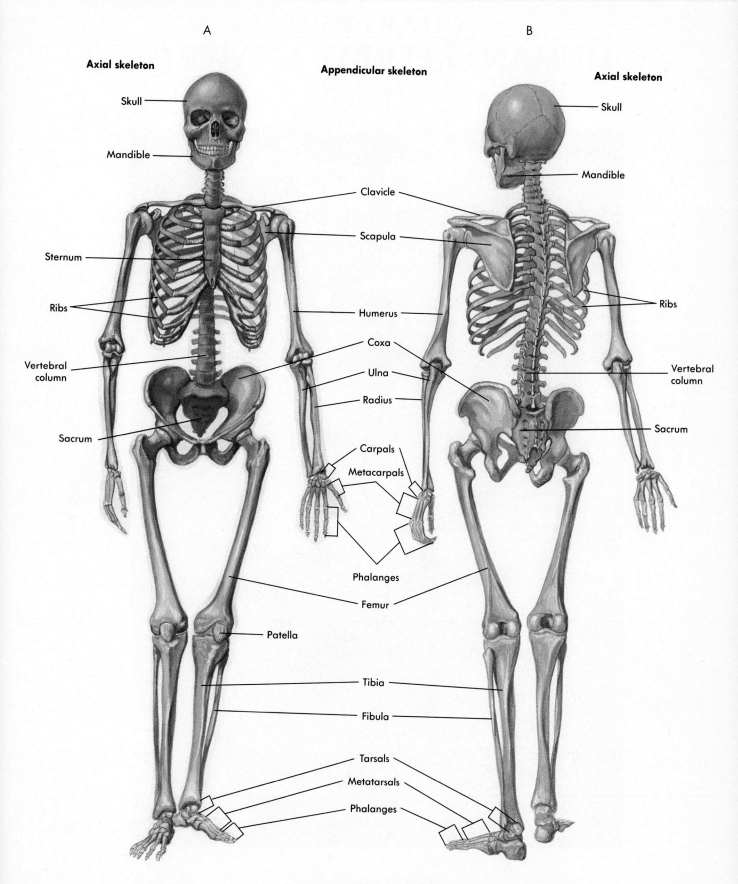

Skull

Mandible

Sternum

Ribs

Vertebral
column

Sacrum

Clavicle

Scapula

Humerus

Coxa

Ulna

Radius

Carpals

Metacarpals

Phalanges

Femur

Patella

Tibia

Fibula

Tarsals

Metatarsals

Phalanges

Skull

Mandible

Ribs

Vertebral
column

Sacrum

Figure 2-1   The Human Skeleton
A, Anterior view. B, Posterior view.

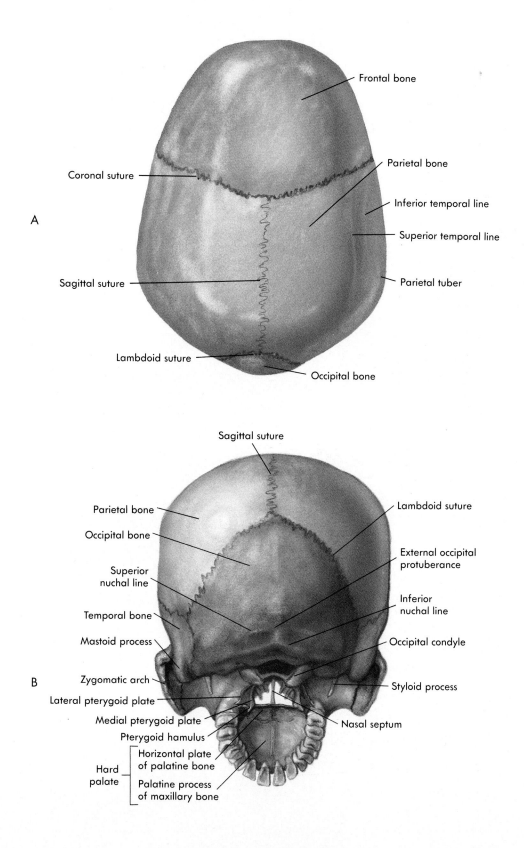

Figure 2-2 The Human Skull
A, Superior view. **B,** Posterior view.

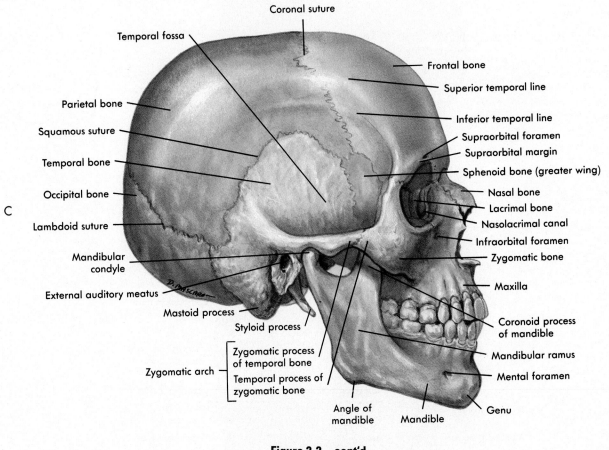

Coronal suture

Temporal fossa

Frontal bone

Superior temporal line

Parietal bone

Inferior temporal line

Squamous suture

Supraorbital foramen

Temporal bone

Supraorbital margin

Occipital bone

Sphenoid bone (greater wing)

Lambdoid suture

Nasal bone

Lacrimal bone

Nasolacrimal canal

Mandibular condyle

Infraorbital foramen

External auditory meatus

Zygomatic bone

Mastoid process

Maxilla

Styloid process

Coronoid process of mandible

Zygomatic process of temporal bone

Temporal process of zygomatic bone

Mandibular ramus

Zygomatic arch

Mental foramen

Angle of mandible

Mandible

Genu

Figure 2-2—cont'd.
C, Lateral view.

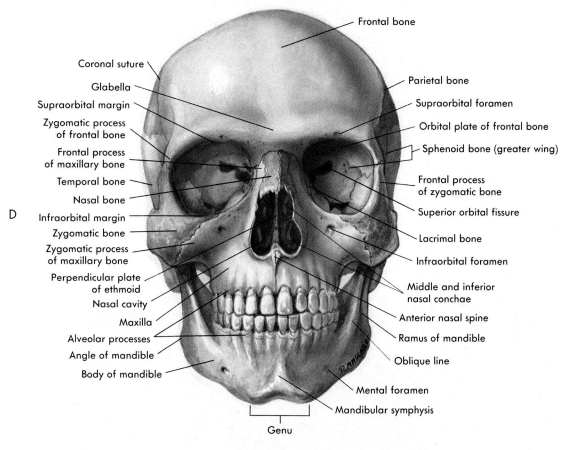

Frontal bone

Coronal suture

Glabella

Supraorbital margin

Zygomatic process
of frontal bone

Frontal process
of maxillary bone

Temporal bone

Nasal bone

Infraorbital margin

Zygomatic bone

Zygomatic process
of maxillary bone

Perpendicular plate
of ethmoid

Nasal cavity

Maxilla

Alveolar processes

Angle of mandible

Body of mandible

Parietal bone

Supraorbital foramen

Orbital plate of frontal bone

Sphenoid bone (greater wing)

Frontal process
of zygomatic bone

Superior orbital fissure

Lacrimal bone

Infraorbital foramen

Middle and inferior
nasal conchae

Anterior nasal spine

Ramus of mandible

Oblique line

Mental foramen

Mandibular symphysis

Genu

D

Figure 2-2—cont'd.
D, Frontal view.

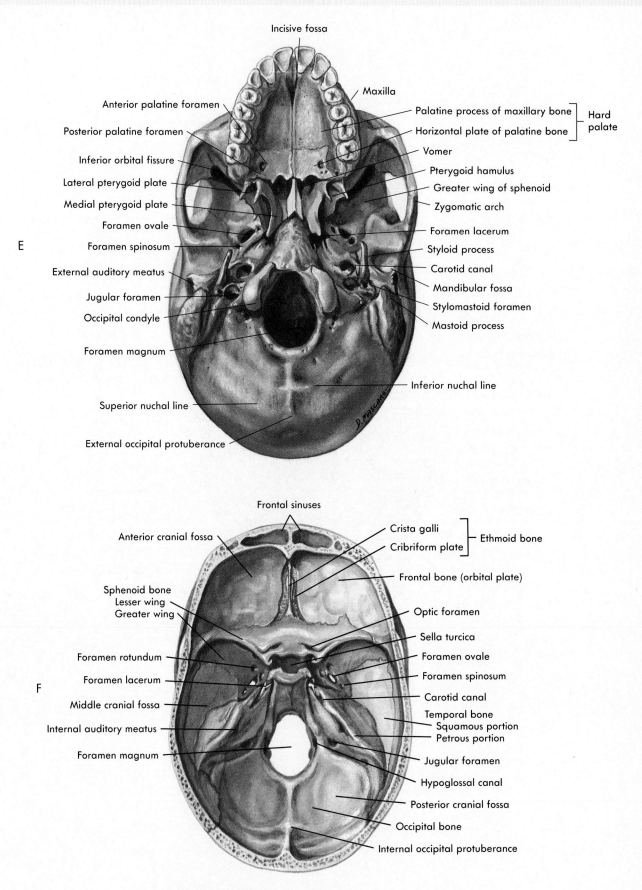

Incisive fossa

Anterior palatine foramen
Posterior palatine foramen
Inferior orbital fissure
Lateral pterygoid plate
Medial pterygoid plate
Foramen ovale
Foramen spinosum
External auditory meatus
Jugular foramen
Occipital condyle

Foramen magnum

Superior nuchal line

External occipital protuberance

Maxilla
Palatine process of maxillary bone ⎤ Hard
Horizontal plate of palatine bone ⎦ palate
Vomer
Pterygoid hamulus
Greater wing of sphenoid
Zygomatic arch
Foramen lacerum
Styloid process
Carotid canal
Mandibular fossa
Stylomastoid foramen
Mastoid process

Inferior nuchal line

E

Frontal sinuses

Anterior cranial fossa

Sphenoid bone
Lesser wing
Greater wing

Foramen rotundum
Foramen lacerum
Middle cranial fossa
Internal auditory meatus
Foramen magnum

Crista galli ⎤ Ethmoid bone
Cribriform plate ⎦
Frontal bone (orbital plate)
Optic foramen
Sella turcica
Foramen ovale
Foramen spinosum
Carotid canal
Temporal bone
Squamous portion
Petrous portion
Jugular foramen
Hypoglossal canal
Posterior cranial fossa
Occipital bone
Internal occipital protuberance

F

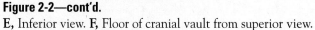

Figure 2-2—cont'd.
E, Inferior view. F, Floor of cranial vault from superior view.

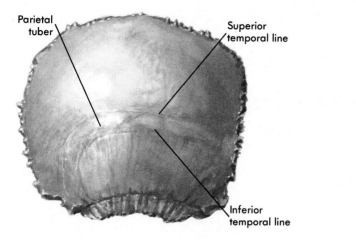

Figure 2-3
Parietal bone, right lateral view.

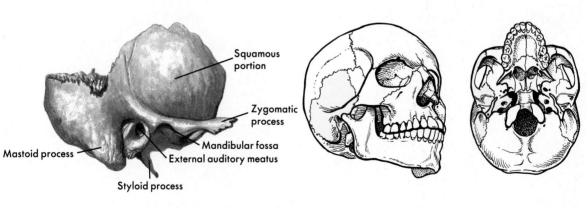

Figure 2-4
Temporal bone, right temporal view.

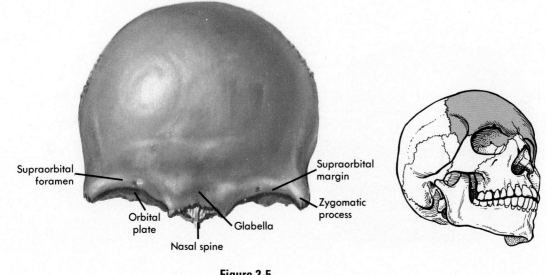

Figure 2-5
Frontal bone, frontal view.

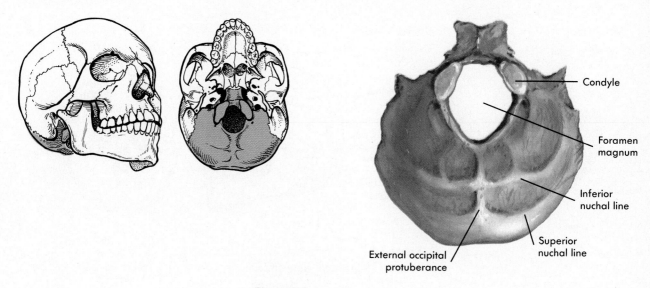

Figure 2-6
Occipital bone, inferior view.

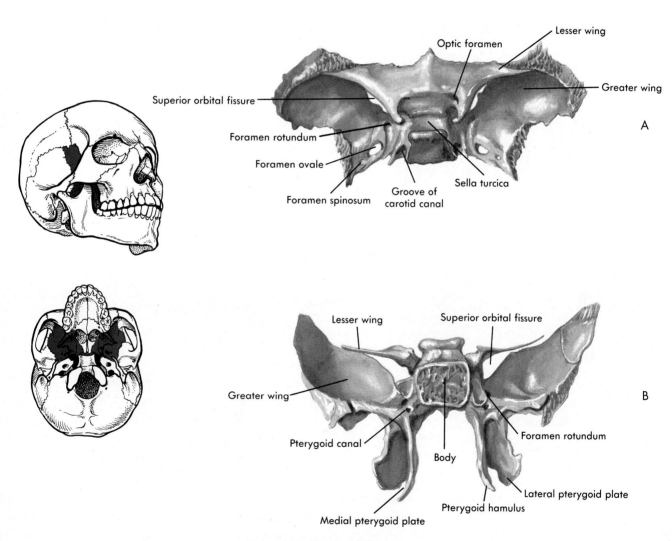

Optic foramen

Lesser wing

Greater wing

Superior orbital fissure

Foramen rotundum

Foramen ovale

Sella turcica

Foramen spinosum

Groove of carotid canal

A

Lesser wing

Superior orbital fissure

Greater wing

Foramen rotundum

Pterygoid canal

Body

Lateral pterygoid plate

Pterygoid hamulus

Medial pterygoid plate

B

Figure 2-7   Sphenoid Bone
A, Superior view. **B,** Posterior view.

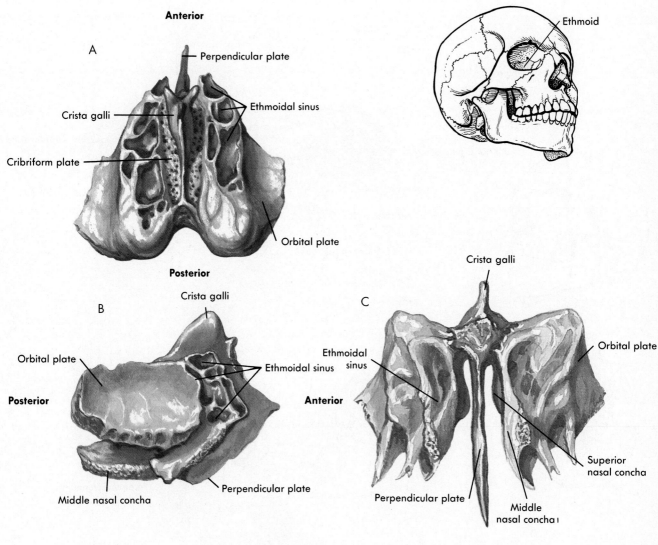

Anterior

A

Perpendicular plate

Ethmoidal sinus

Crista galli

Cribriform plate

Orbital plate

Posterior

Ethmoid

Crista galli

B

Orbital plate

Ethmoidal sinus

Posterior

Middle nasal concha

Perpendicular plate

C

Crista galli

Ethmoidal sinus

Orbital plate

Anterior

Superior nasal concha

Perpendicular plate

Middle nasal concha

Figure 2-8   Ethmoid Bone
A, Superior view. B, Lateral view. C, Anterior view.

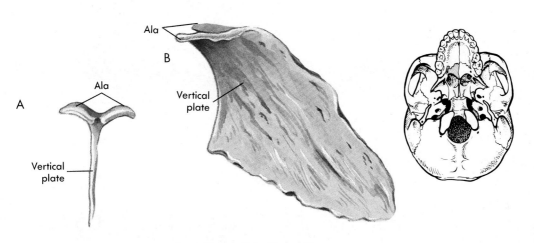

Figure 2-9   Vomer Bone
A, Anterior view. **B,** Lateral view.

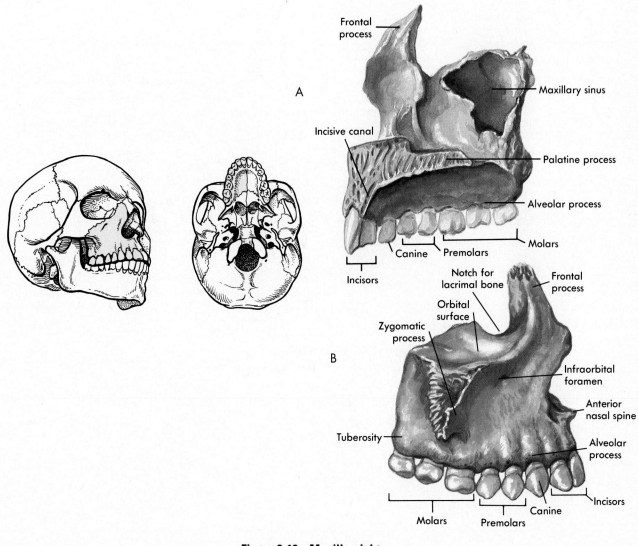

Figure 2-10   Maxilla, right
A, Medial view. **B,** Lateral view.

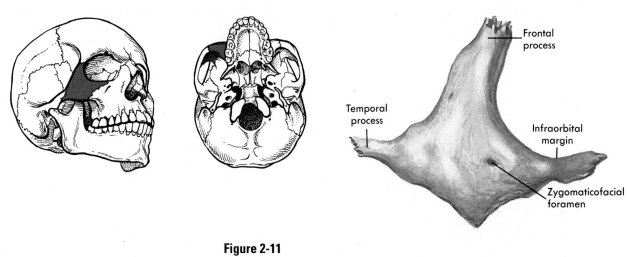

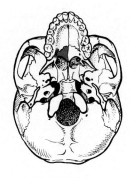

Figure 2-11
Zygomatic bone, right lateral view.

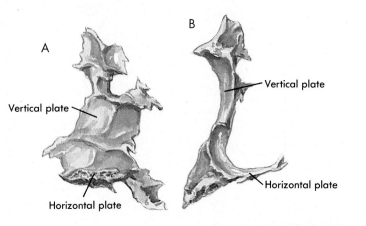

Figure 2-12   Palatine Bone, right
A, Medial view. **B,** Anterior view.

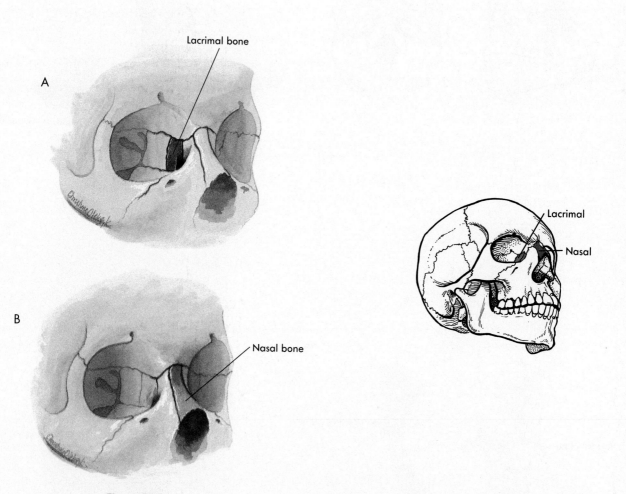

Figure 2-13
A, Lacrimal bone, right anterolateral view. **B,** Nasal bone, right anterolateral view.

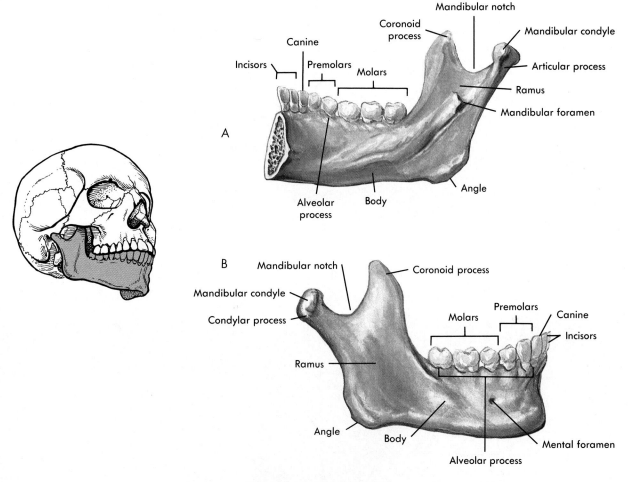

Figure 2-14   Mandible, right
A, Medial view. **B,** Lateral view.

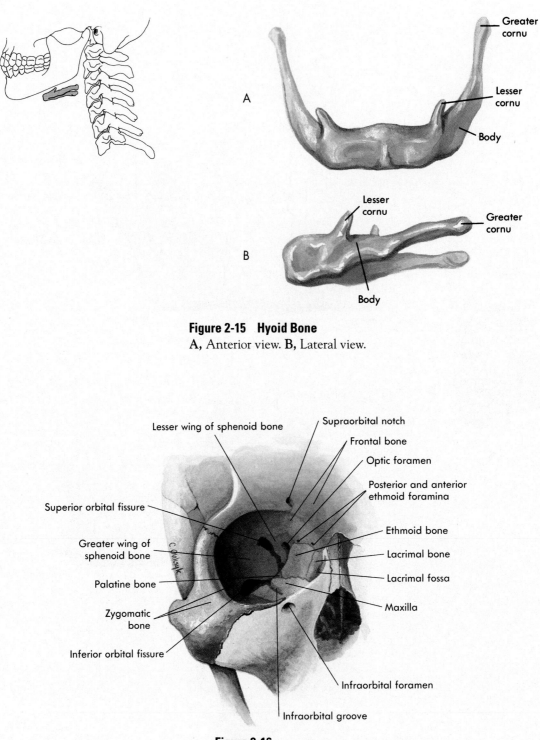

A

B

Greater cornu

Lesser cornu

Body

Lesser cornu

Greater cornu

Body

Figure 2-15 Hyoid Bone
A, Anterior view. **B,** Lateral view.

Lesser wing of sphenoid bone

Supraorbital notch

Frontal bone

Optic foramen

Posterior and anterior ethmoid foramina

Superior orbital fissure

Greater wing of sphenoid bone

Palatine bone

Zygomatic bone

Inferior orbital fissure

Ethmoid bone

Lacrimal bone

Lacrimal fossa

Maxilla

Infraorbital foramen

Infraorbital groove

Figure 2-16
Right orbit, frontal view.

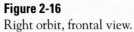

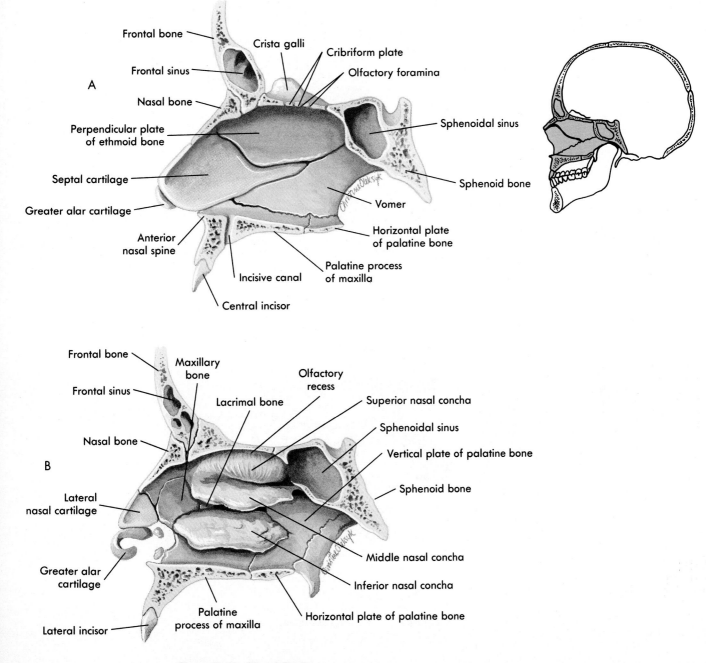

Figure 2-17 Nasal Cavity
A, Nasal septum. B, Right lateral nasal wall, nasal septum removed.

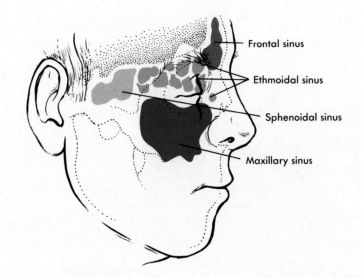

Frontal sinus

Ethmoidal sinus

Sphenoidal sinus

Maxillary sinus

A

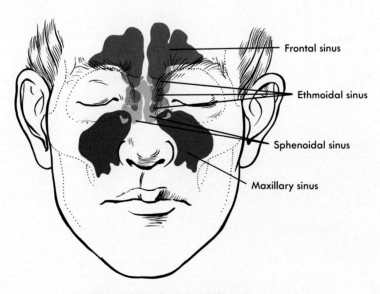

Frontal sinus

Ethmoidal sinus

Sphenoidal sinus

Maxillary sinus

B

Figure 2-18   Paranasal Sinuses
A, Lateral view. **B,** Frontal view.

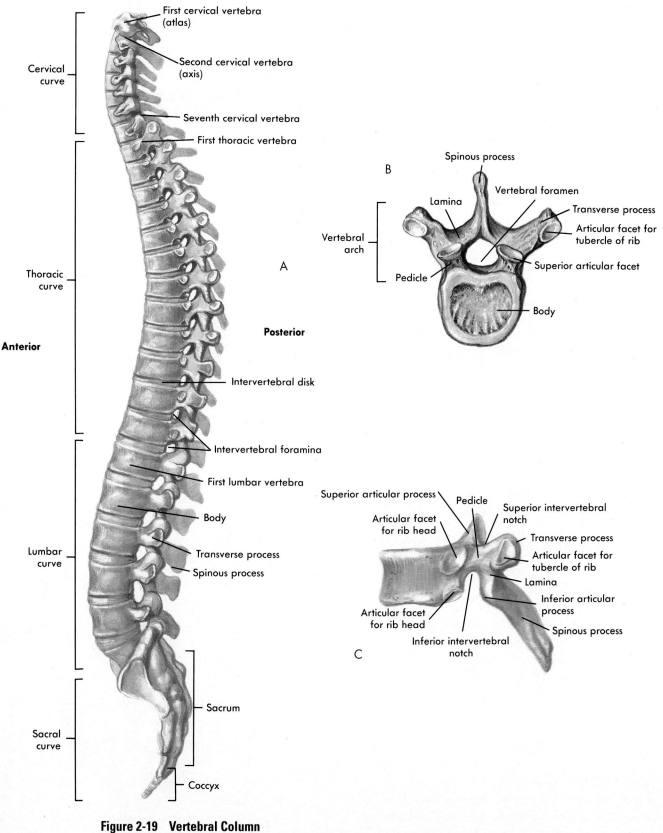

Cervical curve

First cervical vertebra (atlas)

Second cervical vertebra (axis)

Seventh cervical vertebra

First thoracic vertebra

Thoracic curve

A

Anterior

Posterior

Intervertebral disk

Intervertebral foramina

First lumbar vertebra

Body

Lumbar curve

Transverse process

Spinous process

Sacrum

Sacral curve

Coccyx

B

Spinous process

Vertebral foramen

Lamina

Transverse process

Vertebral arch

Articular facet for tubercle of rib

Superior articular facet

Pedicle

Body

Superior articular process

Pedicle

Articular facet for rib head

Superior intervertebral notch

Transverse process

Articular facet for tubercle of rib

Lamina

Inferior articular process

Articular facet for rib head

Spinous process

Inferior intervertebral notch

C

Figure 2-19   Vertebral Column
A, Spinal column, left lateral view. **B,** Vertebra, superior view. **C,** Vertebra, lateral view.

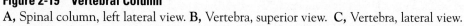

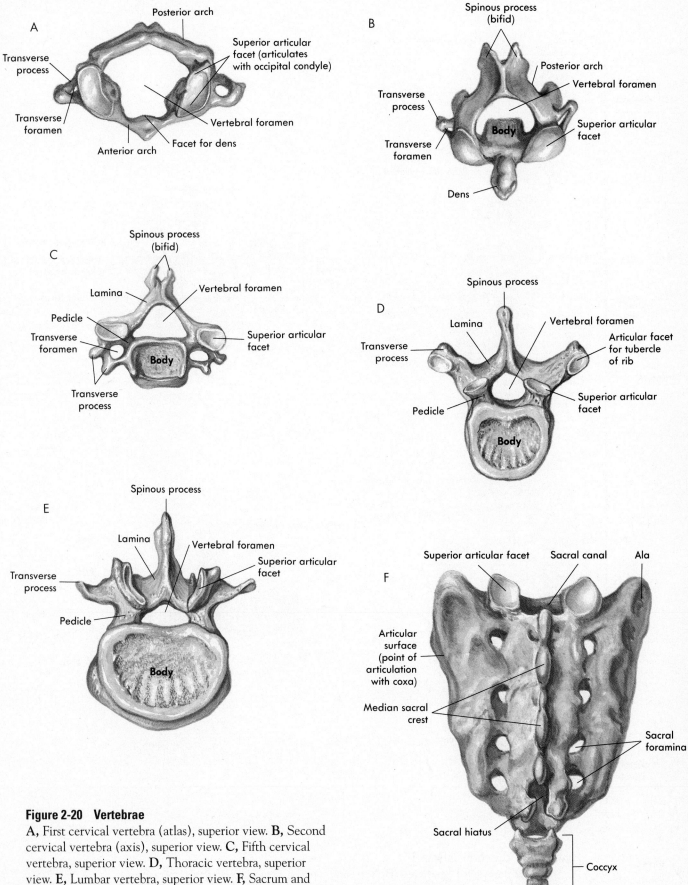

Figure 2-20   Vertebrae

A, First cervical vertebra (atlas), superior view. B, Second cervical vertebra (axis), superior view. C, Fifth cervical vertebra, superior view. D, Thoracic vertebra, superior view. E, Lumbar vertebra, superior view. F, Sacrum and coccyx, posterior view.

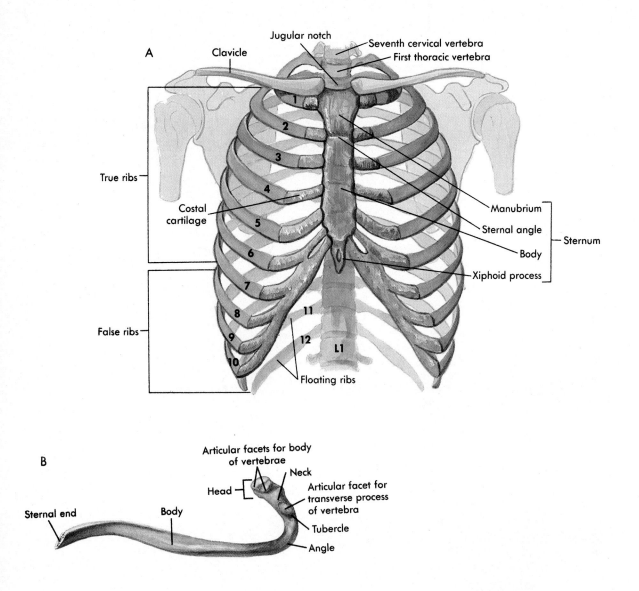

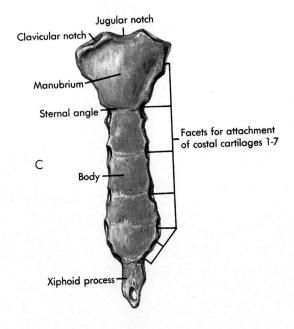

Figure 2-21 Thoracic Cage
A, Rib cage, frontal view. B, Typical rib. C, Sternum, anterior view.

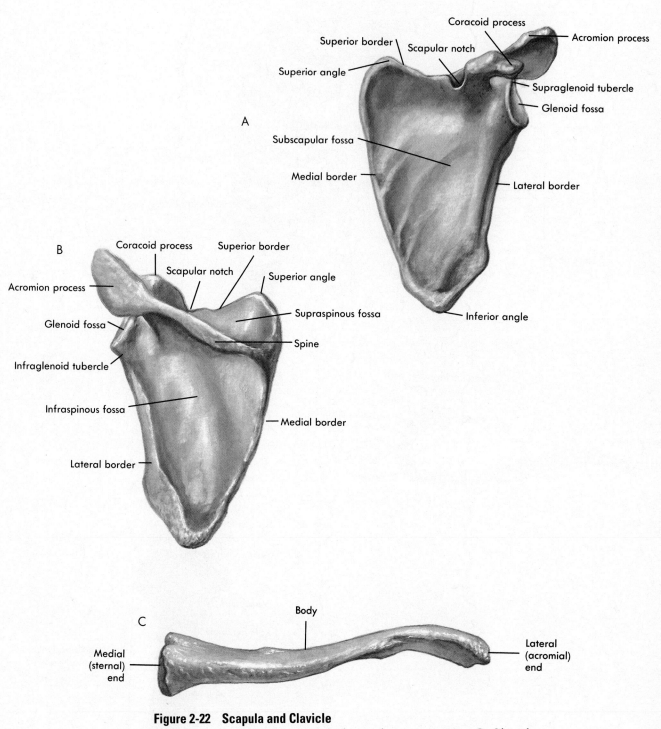

Figure 2-22   Scapula and Clavicle
A, Left scapula, anterior view. B, Left scapula, posterior view. C, Clavicle.

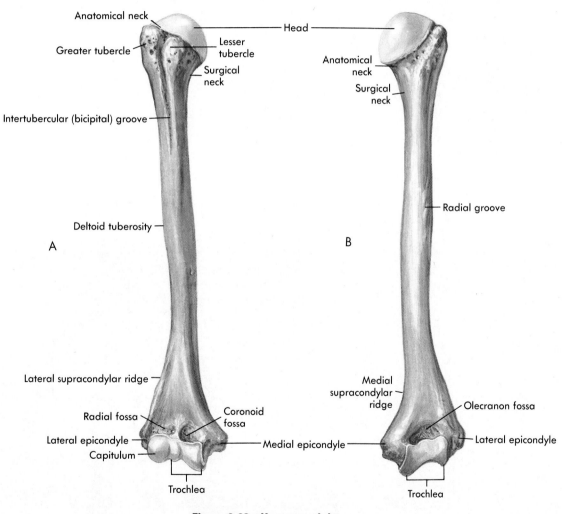

Anatomical neck

Greater tubercle

Lesser tubercle

Head

Anatomical neck

Surgical neck

Surgical neck

Intertubercular (bicipital) groove

Radial groove

Deltoid tuberosity

A

B

Lateral supracondylar ridge

Medial supracondylar ridge

Radial fossa

Coronoid fossa

Olecranon fossa

Lateral epicondyle

Capitulum

Medial epicondyle

Lateral epicondyle

Trochlea

Trochlea

Figure 2-23   Humerus, right
A, Anterior view. **B,** Posterior view.

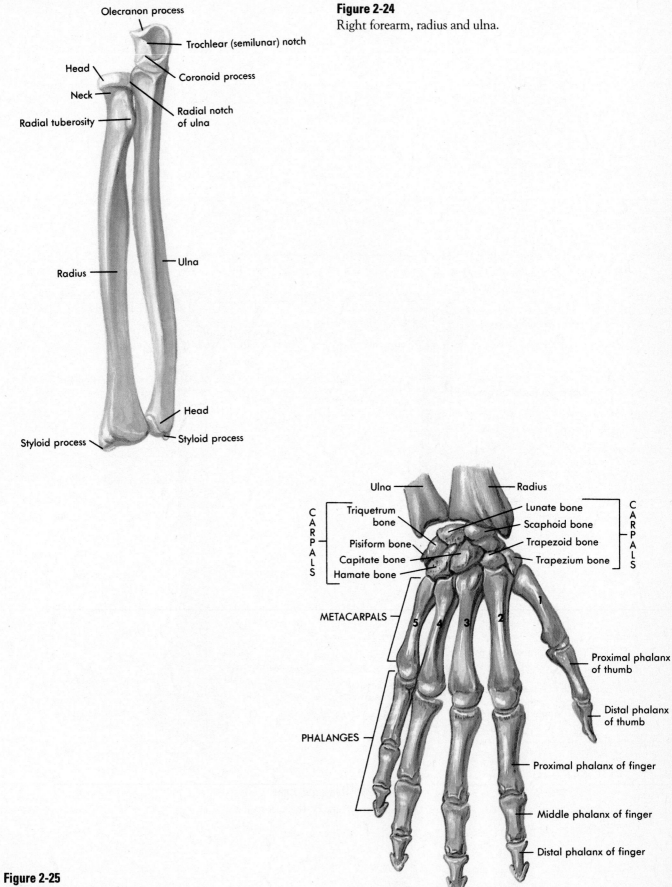

Olecranon process

Trochlear (semilunar) notch

Head

Coronoid process

Neck

Radial tuberosity

Radial notch
of ulna

Radius

Ulna

Head

Styloid process

Styloid process

Figure 2-24
Right forearm, radius and ulna.

Ulna

Radius

C A R P A L S

Triquetrum
bone

Lunate bone

Scaphoid bone

Pisiform bone

Trapezoid bone

Capitate bone

Trapezium bone

Hamate bone

C A R P A L S

METACARPALS

Proximal phalanx
of thumb

Distal phalanx
of thumb

PHALANGES

Proximal phalanx of finger

Middle phalanx of finger

Distal phalanx of finger

Figure 2-25
Wrist and hand, right posterior view.

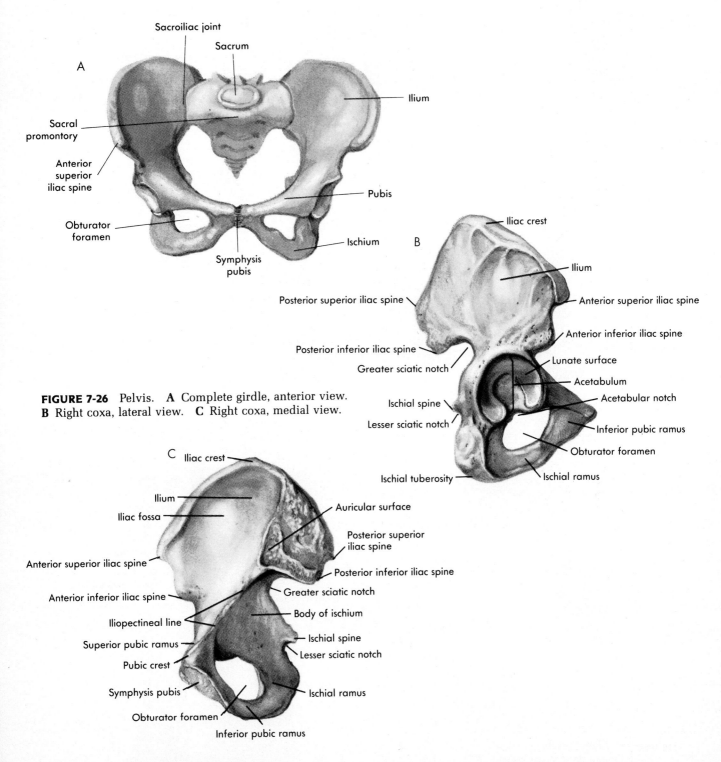

A

Sacroiliac joint

Sacrum

Sacral promontory

Anterior superior iliac spine

Obturator foramen

Symphysis pubis

Ilium

Pubis

Ischium

B

Iliac crest

Ilium

Posterior superior iliac spine

Anterior superior iliac spine

Anterior inferior iliac spine

Posterior inferior iliac spine

Greater sciatic notch

Lunate surface

Acetabulum

Acetabular notch

Ischial spine

Lesser sciatic notch

Inferior pubic ramus

Obturator foramen

Ischial tuberosity

Ischial ramus

FIGURE 7-26 Pelvis. **A** Complete girdle, anterior view.
B Right coxa, lateral view. **C** Right coxa, medial view.

C

Iliac crest

Ilium

Iliac fossa

Anterior superior iliac spine

Anterior inferior iliac spine

Iliopectineal line

Superior pubic ramus

Pubic crest

Symphysis pubis

Obturator foramen

Inferior pubic ramus

Auricular surface

Posterior superior iliac spine

Posterior inferior iliac spine

Greater sciatic notch

Body of ischium

Ischial spine

Lesser sciatic notch

Ischial ramus

Figure 2-26   Pelvis
A, Anterior view. **B,** Right coxa, lateral view. **C,** Right coxa, medial view.

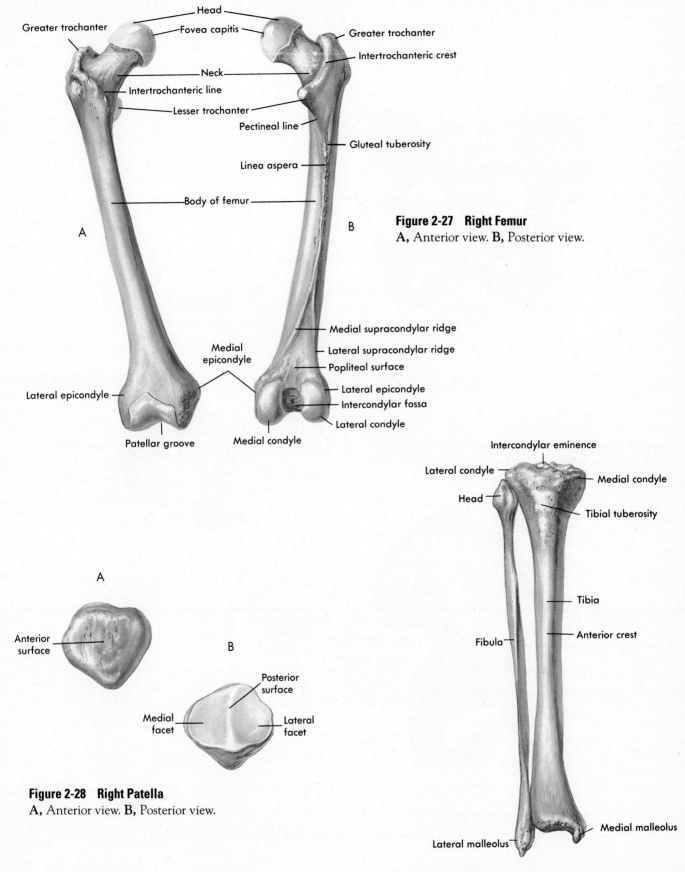

Greater trochanter
Head
Fovea capitis
Greater trochanter
Intertrochanteric crest
Neck
Intertrochanteric line
Lesser trochanter
Pectineal line
Gluteal tuberosity
Linea aspera
Body of femur

A

B

Figure 2-27   Right Femur
A, Anterior view. **B,** Posterior view.

Medial supracondylar ridge
Lateral supracondylar ridge
Popliteal surface
Lateral epicondyle
Intercondylar fossa
Lateral condyle

Medial epicondyle
Lateral epicondyle
Medial condyle
Patellar groove

Intercondylar eminence
Lateral condyle
Medial condyle
Head
Tibial tuberosity

A

Anterior surface

B

Posterior surface
Medial facet
Lateral facet

Tibia
Anterior crest
Fibula

Figure 2-28   Right Patella
A, Anterior view. **B,** Posterior view.

Medial malleolus
Lateral malleolus

Figure 2-29
Tibia and fibula, right anterior view.

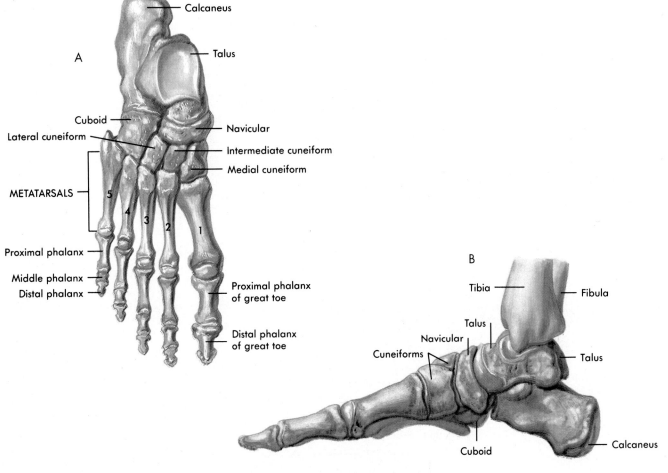

Figure 2-30   Right Ankle and Foot
A, Dorsal view. B, Medial view.

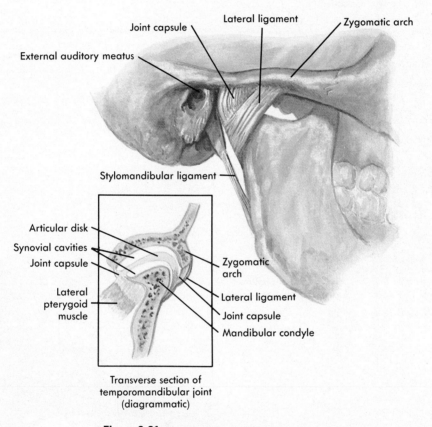

Joint capsule
Lateral ligament
Zygomatic arch
External auditory meatus
Stylomandibular ligament

Articular disk
Synovial cavities
Joint capsule
Lateral pterygoid muscle
Zygomatic arch
Lateral ligament
Joint capsule
Mandibular condyle

Transverse section of
temporomandibular joint
(diagrammatic)

Figure 2-31
Right temporomandibular joint, lateral view.

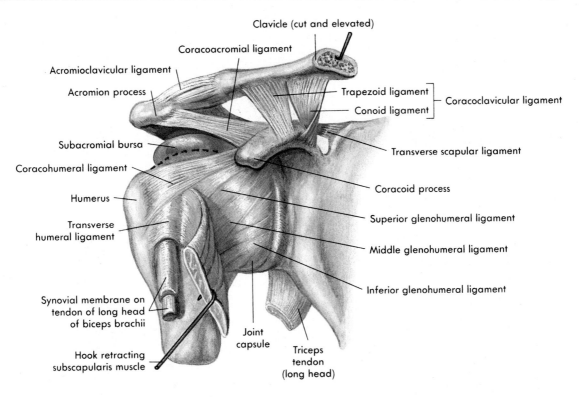

Clavicle (cut and elevated)

Coracoacromial ligament

Acromioclavicular ligament

Acromion process

Trapezoid ligament

Conoid ligament

Coracoclavicular ligament

Subacromial bursa

Coracohumeral ligament

Humerus

Transverse scapular ligament

Transverse humeral ligament

Coracoid process

Superior glenohumeral ligament

Middle glenohumeral ligament

Inferior glenohumeral ligament

Synovial membrane on tendon of long head of biceps brachii

Hook retracting subscapularis muscle

Joint capsule

Triceps tendon (long head)

Figure 2-32
Ligaments of the right shoulder joint, anterior view.

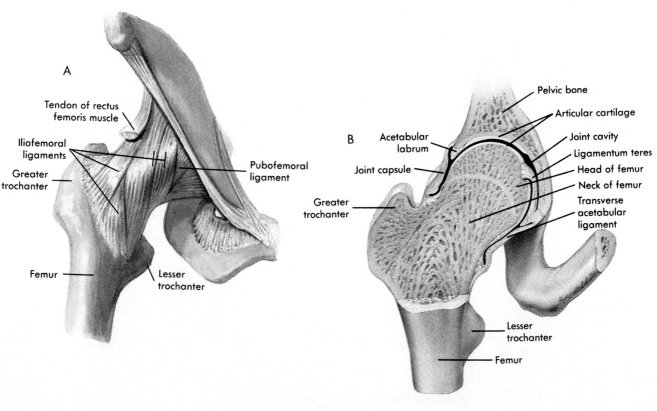

A

Tendon of rectus femoris muscle

Iliofemoral ligaments

Greater trochanter

Pubofemoral ligament

Femur

Lesser trochanter

B

Acetabular labrum

Joint capsule

Greater trochanter

Pelvic bone

Articular cartilage

Joint cavity

Ligamentum teres

Head of femur

Neck of femur

Transverse acetabular ligament

Lesser trochanter

Femur

Figure 2-33   Right Hip Joint
A, Anterior view. **B,** Frontal section.

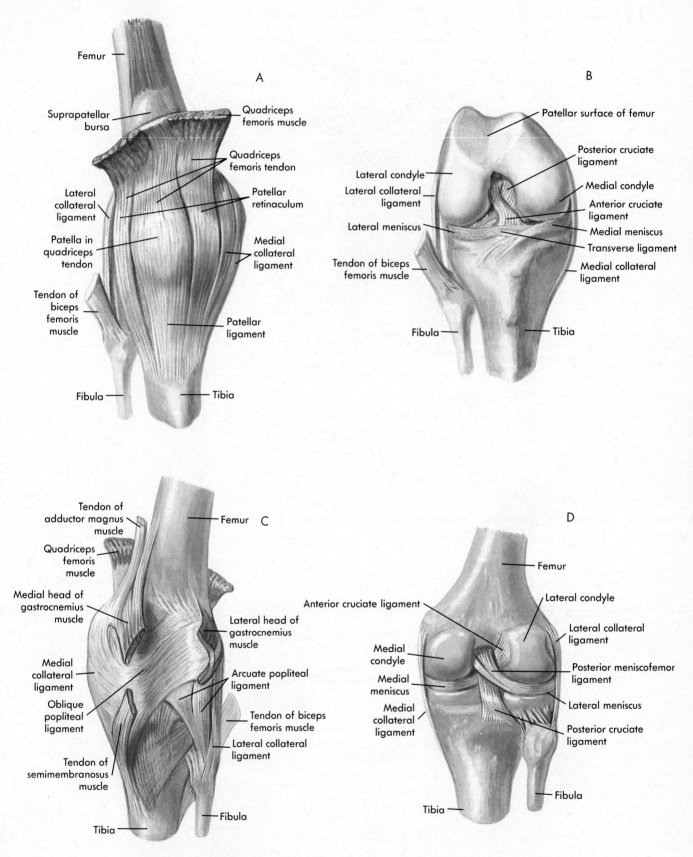

Figure 2-34   Right Knee Joint

A, Anterior superficial view. **B,** Anterior deep view with knee flexed. **C,** Posterior superficial view. **D,** Posterior deep view.

CHAPTER 3
HUMAN MUSCULAR ANATOMY

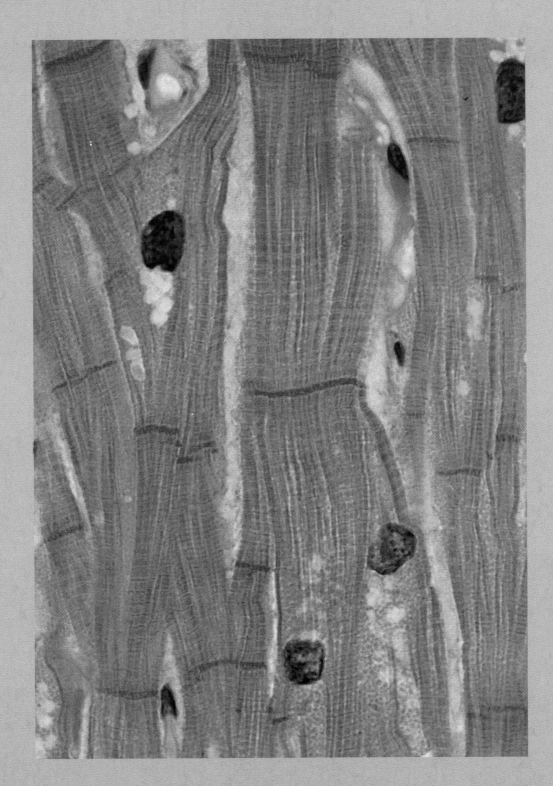

Cardiac Muscle (Longitudinal Section)

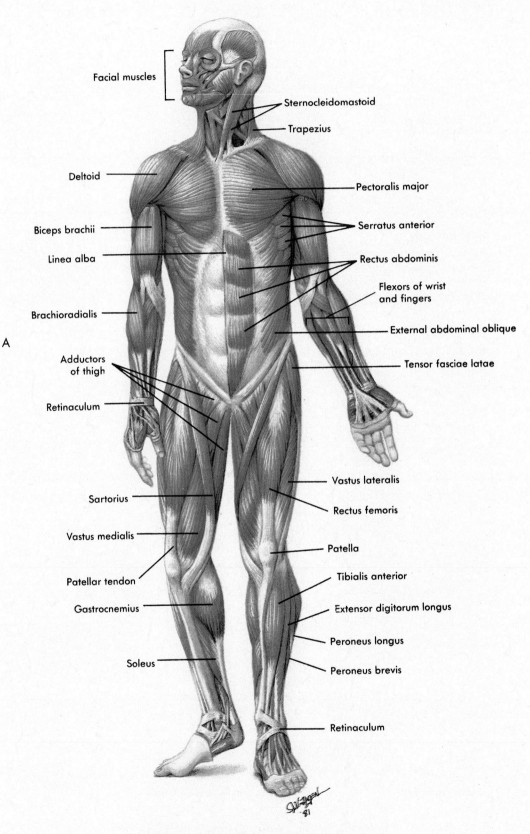

Facial muscles

Sternocleidomastoid

Trapezius

Deltoid

Pectoralis major

Biceps brachii

Serratus anterior

Linea alba

Rectus abdominis

Flexors of wrist
and fingers

Brachioradialis

External abdominal oblique

A

Tensor fasciae latae

Adductors
of thigh

Retinaculum

Vastus lateralis

Sartorius

Rectus femoris

Vastus medialis

Patella

Patellar tendon

Tibialis anterior

Gastrocnemius

Extensor digitorum longus

Peroneus longus

Peroneus brevis

Soleus

Retinaculum

Figure 3-1 Major Muscles of the Human Body
A, Anterior view.

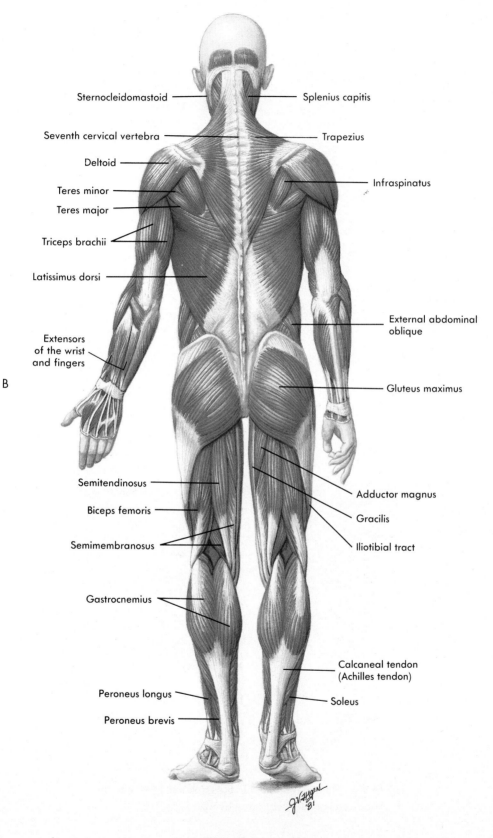

Sternocleidomastoid

Seventh cervical vertebra

Deltoid

Teres minor

Teres major

Triceps brachii

Latissimus dorsi

Extensors
of the wrist
and fingers

B

Splenius capitis

Trapezius

Infraspinatus

External abdominal
oblique

Gluteus maximus

Semitendinosus

Biceps femoris

Semimembranosus

Gastrocnemius

Adductor magnus

Gracilis

Iliotibial tract

Peroneus longus

Peroneus brevis

Calcaneal tendon
(Achilles tendon)

Soleus

Figure 3-1—cont'd.
B, Posterior view.

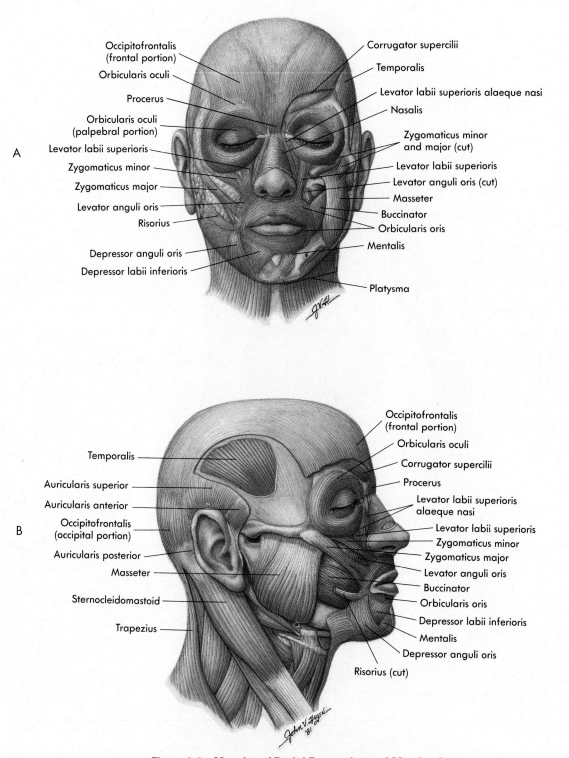

Figure 3-2   Muscles of Facial Expression and Mastication
A, Anterior view. **B,** Right lateral view.

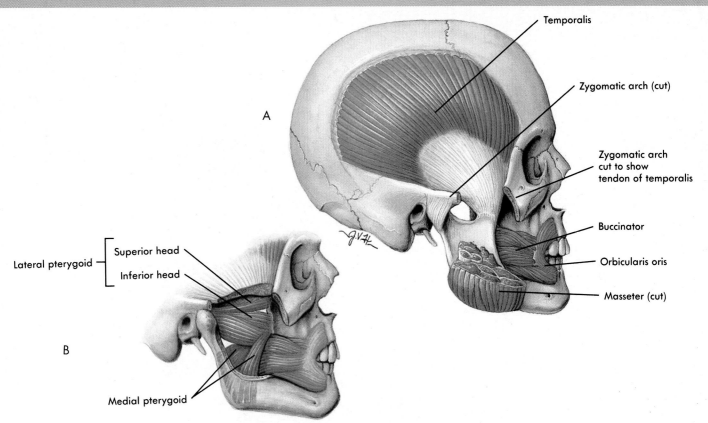

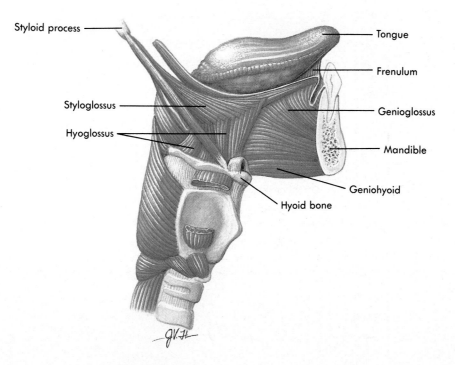

Figure 3-3   Muscles of Mastication

A, Right lateral view. The masseter muscle has been cut and the zygomatic arch has been removed to expose the temporalis muscle. **B,** Deep view. The temporalis and masseter muscles have been cut and the zygomatic arch and part of the mandible have been removed.

Figure 3-4
Muscles of the tongue from a right lateral view.

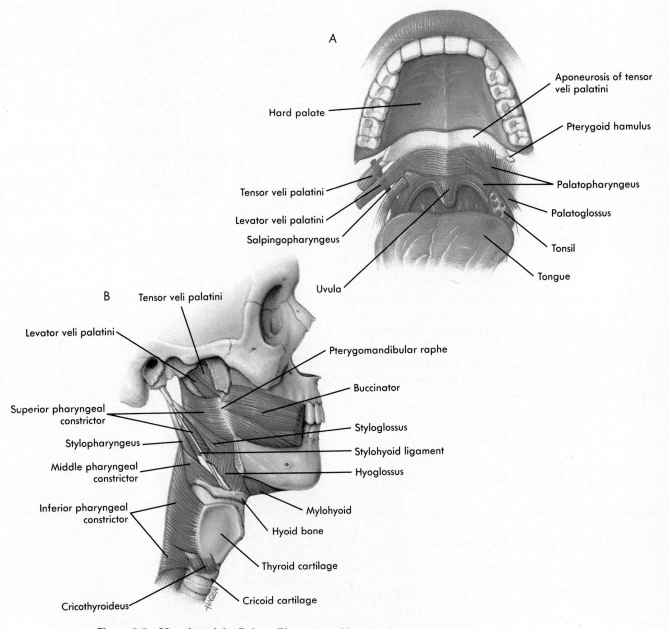

Figure 3-5   Muscles of the Palate, Pharynx, and Larynx
A, Inferior view of the palate. The palatoglossus and part of the palatopharyngeus muscles have been cut on the right side to reveal the deeper muscles. **B,** Right lateral view of the pharynx and larynx. Part of the mandible has been removed to reveal deeper structures.

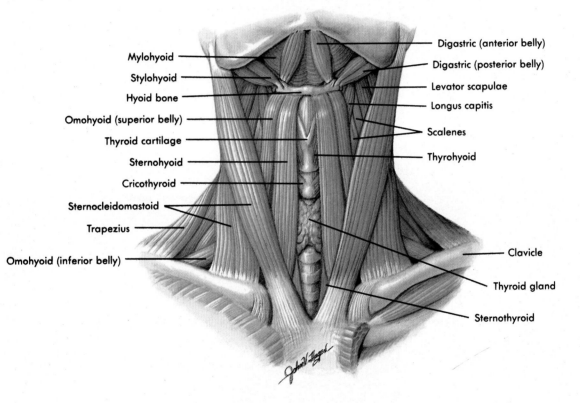

Mylohyoid

Stylohyoid

Hyoid bone

Omohyoid (superior belly)

Thyroid cartilage

Sternohyoid

Cricothyroid

Sternocleidomastoid

Trapezius

Omohyoid (inferior belly)

Digastric (anterior belly)

Digastric (posterior belly)

Levator scapulae

Longus capitis

Scalenes

Thyrohyoid

Clavicle

Thyroid gland

Sternothyroid

Figure 3-6
Muscles of the anterior neck, superficial view.

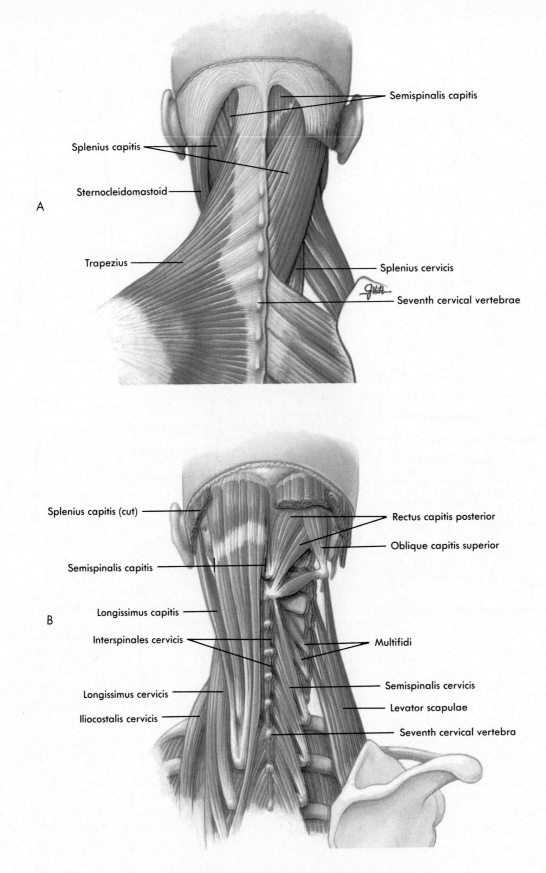

A,

- Semispinalis capitis
- Splenius capitis
- Sternocleidomastoid
- Trapezius
- Splenius cervicis
- Seventh cervical vertebrae

B,

- Splenius capitis (cut)
- Semispinalis capitis
- Longissimus capitis
- Interspinales cervicis
- Longissimus cervicis
- Iliocostalis cervicis
- Rectus capitis posterior
- Oblique capitis superior
- Multifidi
- Semispinalis cervicis
- Levator scapulae
- Seventh cervical vertebra

Figure 3-7   Muscles of the Posterior Neck
A, Superficial view. **B,** Deep view.

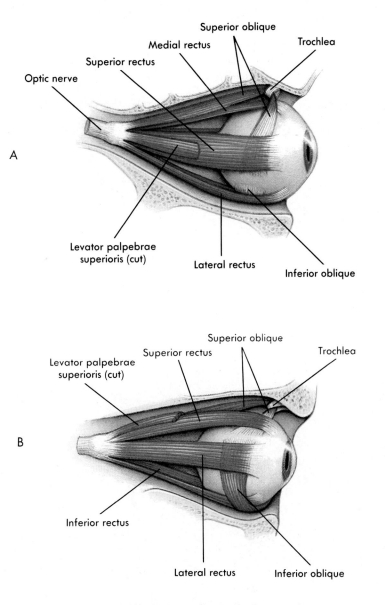

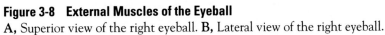

Figure 3-8   External Muscles of the Eyeball

A, Superior view of the right eyeball. **B,** Lateral view of the right eyeball.

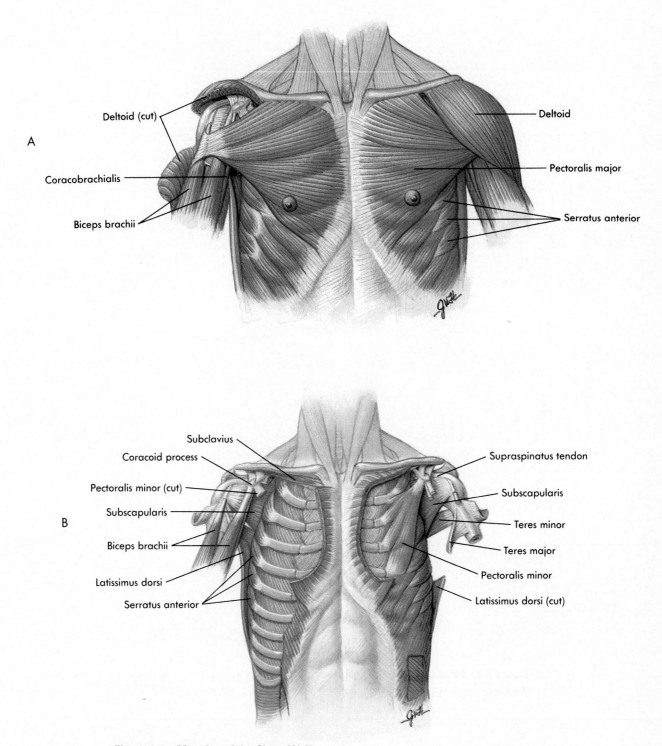

Figure 3-9   Muscles of the Chest Wall

A, Anterior superficial view. **B,** Anterior deep view. The pectoralis major muscles have been removed on both sides. The pectoralis minor muscles has been removed on the right side.

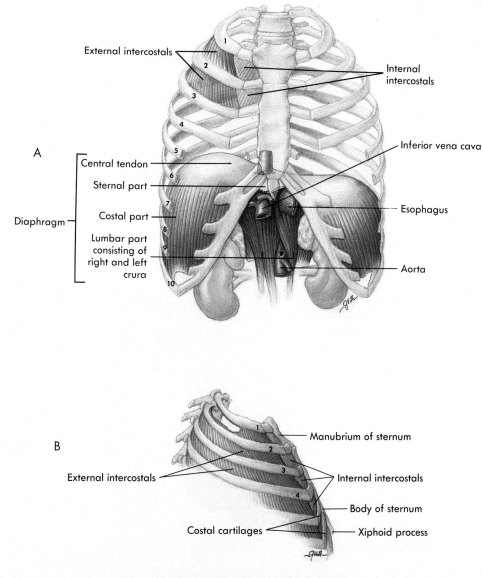

Figure 3-10   Muscles of the Thorax
A, Anterior view. **B,** Right lateral view.

A

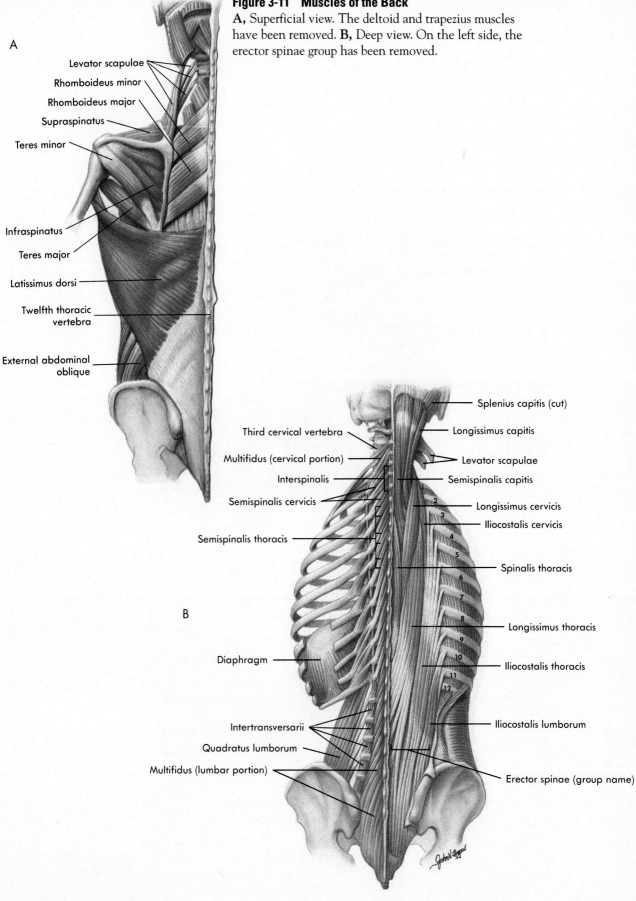

Figure 3-11   Muscles of the Back

A, Superficial view. The deltoid and trapezius muscles have been removed. **B,** Deep view. On the left side, the erector spinae group has been removed.

Levator scapulae
Rhomboideus minor
Rhomboideus major
Supraspinatus
Teres minor
Infraspinatus
Teres major
Latissimus dorsi
Twelfth thoracic vertebra
External abdominal oblique

Splenius capitis (cut)
Longissimus capitis
Third cervical vertebra
Multifidus (cervical portion)
Levator scapulae
Interspinalis
Semispinalis capitis
Semispinalis cervicis
Longissimus cervicis
Iliocostalis cervicis
Semispinalis thoracis
Spinalis thoracis
Longissimus thoracis
Diaphragm
Iliocostalis thoracis
Intertransversarii
Iliocostalis lumborum
Quadratus lumborum
Multifidus (lumbar portion)
Erector spinae (group name)

B

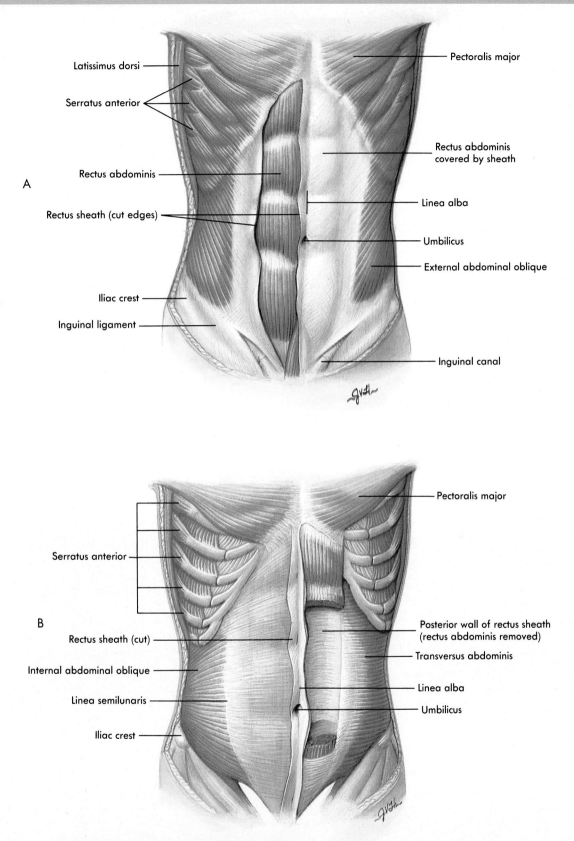

Latissimus dorsi

Serratus anterior

Rectus abdominis

Rectus sheath (cut edges)

Iliac crest

Inguinal ligament

Pectoralis major

Rectus abdominis covered by sheath

Linea alba

Umbilicus

External abdominal oblique

Inguinal canal

A

Serratus anterior

Rectus sheath (cut)

Internal abdominal oblique

Linea semilunaris

Iliac crest

Pectoralis major

Posterior wall of rectus sheath (rectus abdominis removed)

Transversus abdominis

Linea alba

Umbilicus

B

Figure 3-12   Muscles of the Abdominal Wall

A, Superficial view. The rectus abdominis sheath has been removed on the right side.
B, Deep view. On the right side, the external oblique muscle has been removed. On the left side, the external and internal oblique muscles have been removed. Also on the left side, the rectus abdominis muscle has been removed to reveal the posterior rectus sheath.

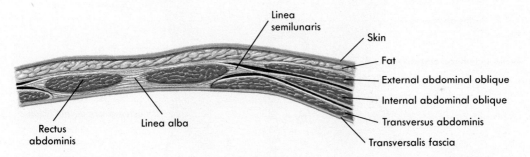
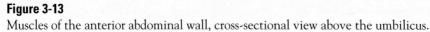

Figure 3-13
Muscles of the anterior abdominal wall, cross-sectional view above the umbilicus.

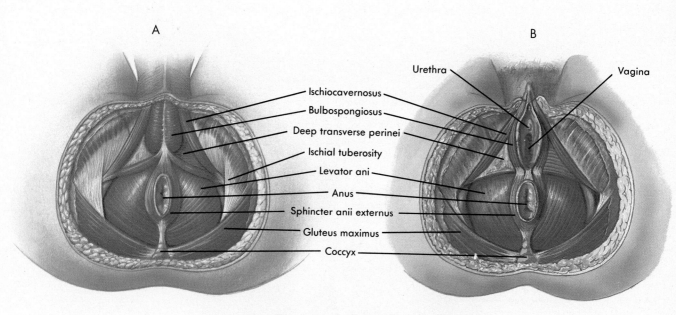
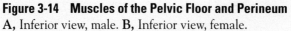

Figure 3-14 Muscles of the Pelvic Floor and Perineum
A, Inferior view, male. **B,** Inferior view, female.

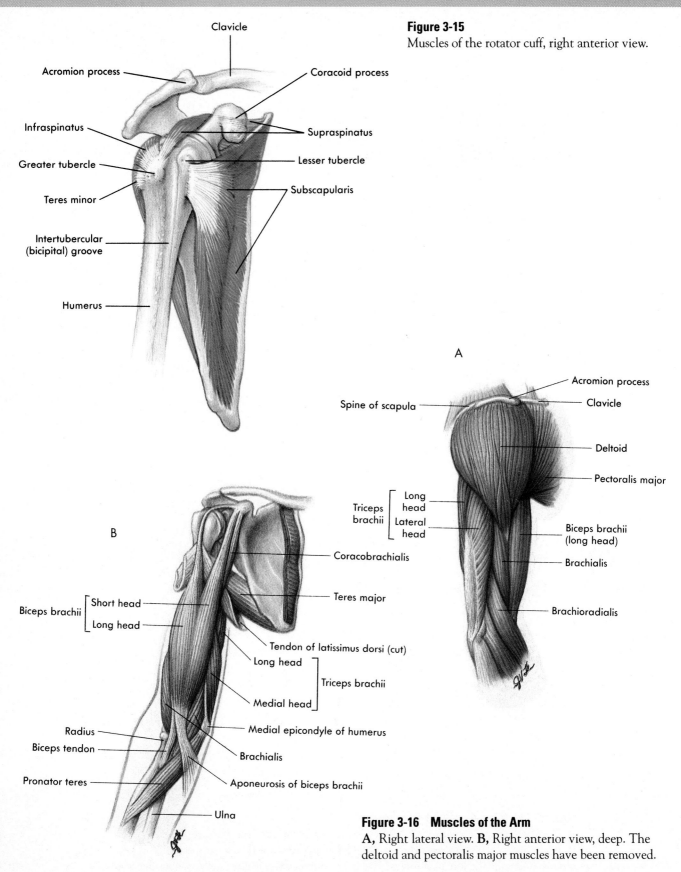

Clavicle

Acromion process

Coracoid process

Infraspinatus

Supraspinatus

Greater tubercle

Lesser tubercle

Teres minor

Subscapularis

Intertubercular (bicipital) groove

Humerus

Figure 3-15
Muscles of the rotator cuff, right anterior view.

A

Spine of scapula

Acromion process

Clavicle

Deltoid

Pectoralis major

Triceps brachii { Long head / Lateral head }

Biceps brachii (long head)

Brachialis

Brachioradialis

B

Coracobrachialis

Biceps brachii { Short head / Long head }

Teres major

Tendon of latissimus dorsi (cut)

Long head

Triceps brachii

Medial head

Radius

Medial epicondyle of humerus

Biceps tendon

Brachialis

Pronator teres

Aponeurosis of biceps brachii

Ulna

Figure 3-16   Muscles of the Arm
A, Right lateral view. **B,** Right anterior view, deep. The deltoid and pectoralis major muscles have been removed.

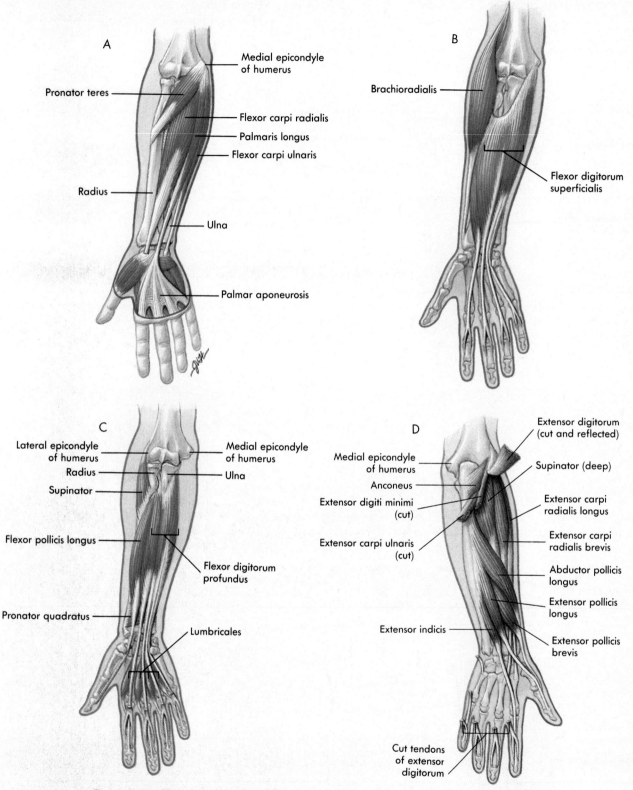

A, Right anterior view, superficial. The brachioradialis muscle has been removed.
B, Right anterior view, deeper than in **A.** The pronator teres, flexor carpi radialis and ulnaris, and palmaris longus muscles have been removed. **C,** Right anterior view, deeper than in **B.** Brachioradialis, pronator teres, flexor carpi radialis and ulnaris, palmaris longus, and flexor digitorum superficialis muscles have been removed. **D,** Right posterior view, deep. Extensor digitorum, extensor digiti minimi, and extensor carpi ulnaris muscles have been removed.

Figure 3-17   Muscles of the Forearm

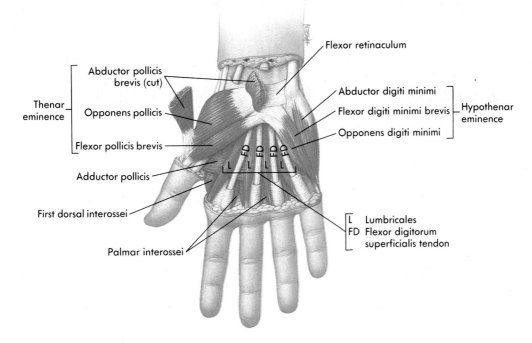

Figure 3-18
Muscles of the right hand, palmar surface.

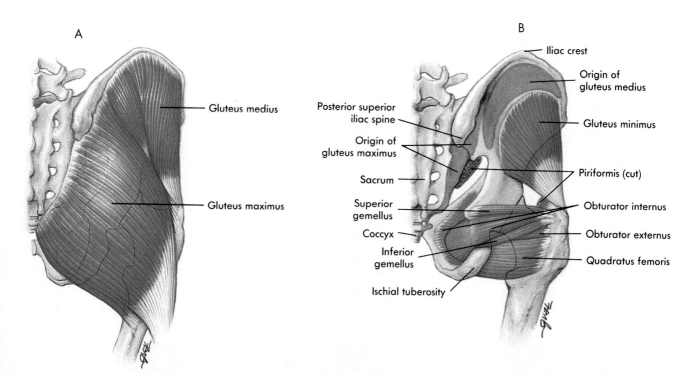

Figure 3-19   Muscles of the Right Posterior Hip
A, Superficial view. **B,** Deep view. The gluteus maximus and gluteus medius muscles have been removed. The piriformis muscle has been cut.

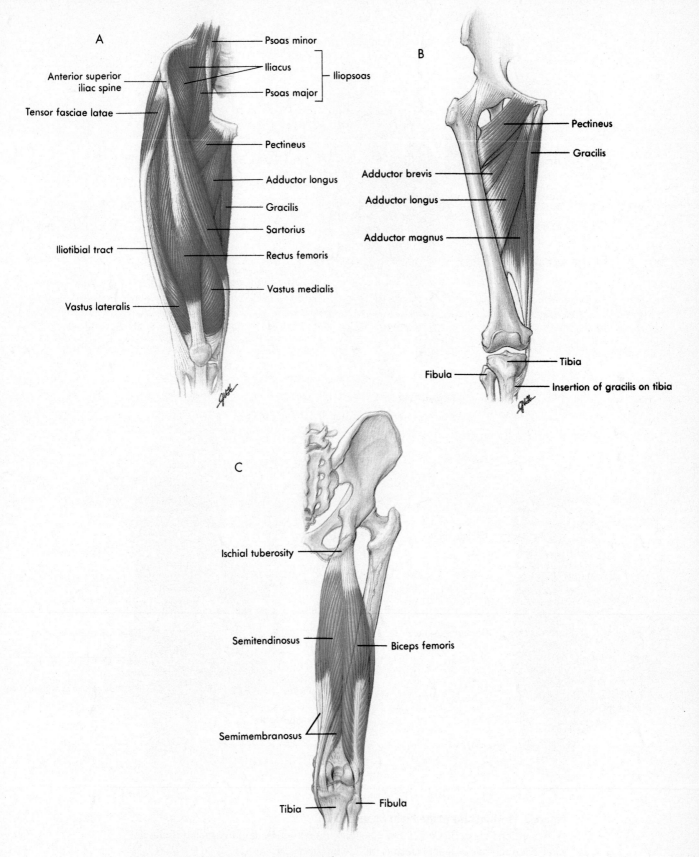

Figure 3-20   Muscles of the Right Thigh
A, Anterior superficial view. B, Adductor region. Tensor fasciae latae, sartorius, and quadriceps muscles have been removed. C, Posterior view. Hip muscles have been removed.

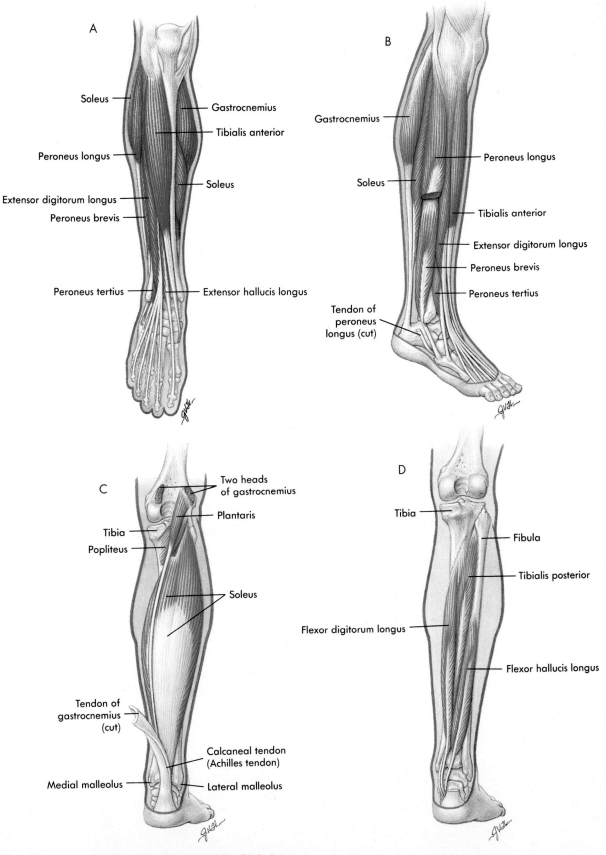

A

Soleus

Gastrocnemius

Tibialis anterior

Peroneus longus

Soleus

Extensor digitorum longus

Peroneus brevis

Peroneus tertius

Extensor hallucis longus

B

Gastrocnemius

Peroneus longus

Soleus

Tibialis anterior

Extensor digitorum longus

Peroneus brevis

Peroneus tertius

Tendon of peroneus longus (cut)

C

Two heads of gastrocnemius

Plantaris

Tibia

Popliteus

Soleus

Tendon of gastrocnemius (cut)

Calcaneal tendon (Achilles tendon)

Medial malleolus

Lateral malleolus

D

Tibia

Fibula

Tibialis posterior

Flexor digitorum longus

Flexor hallucis longus

Figure 3-21   Muscles of the Right Leg
A, Anterior view. **B,** Lateral view. **C,** Posterior superficial view. **D,** Posterior deep view.
The gastrocnemius, plantaris, and soleus muscles have been removed.

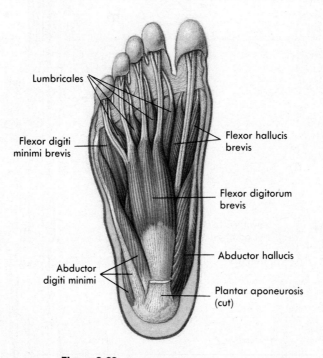

Lumbricales

Flexor digiti
minimi brevis

Flexor hallucis
brevis

Flexor digitorum
brevis

Abductor hallucis

Abductor
digiti minimi

Plantar aponeurosis
(cut)

Figure 3-22
Muscles of the right foot, plantar surface.

CHAPTER 4
DISSECTIONS

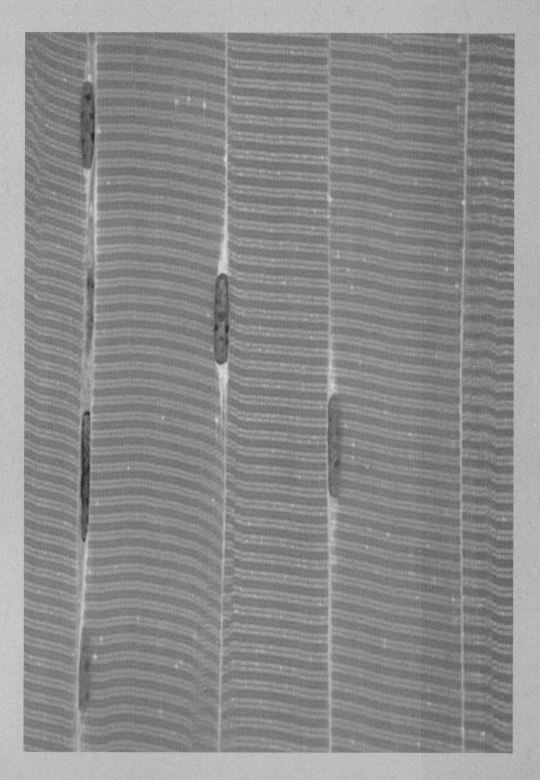

Striated (Skeletal) Muscle Fibers (Longitudinal Section)

Figure 4-1
The Cat Skeleton
1. Maxilla
2. Mandible
3. Orbit
4. Zygomatic arch
5. Cranium
6. Cervical vertebrae (7)
7. Sternum
8. Scapula
9. Humerus
10. Radius
11. Ulna
12. Carpal bones
13. Metacarpal bones
14. Phalanges
15. Thoracic vertebrae (13)
16. Ribs
17. Lumbar vertebrae (7)
18. Pelvis
19. Femur
20. Tibia
21. Fibula
22. Calcaneus
23. Tarsal bones
24. Metatarsal bones
25. Phalanges
26. Caudal vertebrae (21–25)

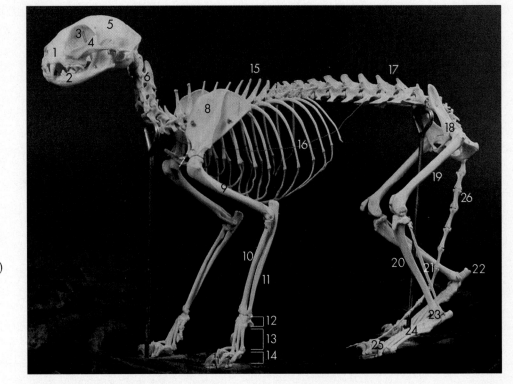

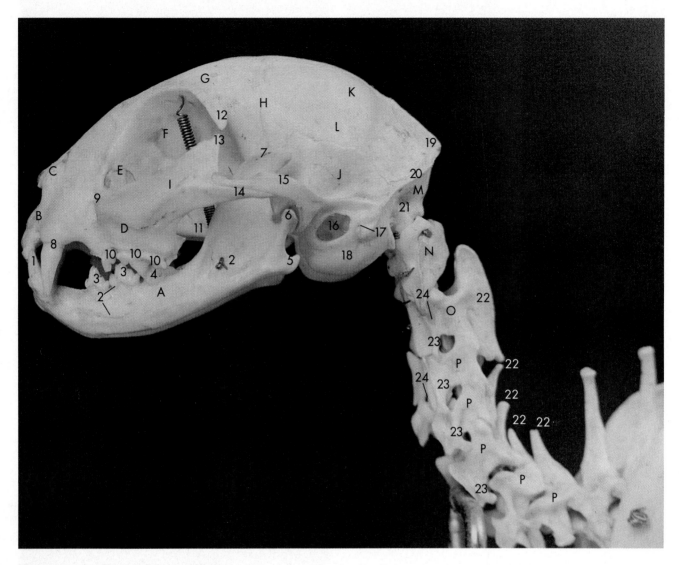

Figure 4-2   Cat Skull, Left Lateral View

A. Mandible
   1. Lower canine tooth
   2. Mental foramina
   3. Lower premolar teeth
   4. Lower molar teeth
   5. Angular process
   6. Condyloid process
   7. Coronoid process
B. Incisive bone
C. Nasal bone
D. Maxilla
   8. Upper canine tooth
   9. Infraorbital foramen
   10. Upper premolar teeth
   11. Upper molar tooth

E. Lacrimal bone and fossa
F. Orbit
G. Frontal bone
   12. Zygomatic process of frontal
      bone
H. Coronal suture
I. Malar or zygomatic bone
   13. Frontal process of malar
   14. Temporal process of malar
J. Temporal bone
   15. Zygomatic process of
      temporal bone
   16. External auditory meatus
   17. Stylomastoid foramen
   18. Mastoid process

K. Parietal bone
L. Squamosal suture
M. Occipital bone
   19. External occipital
      protuberance
   20. Nuchal crest
   21. Occipital condyle
N. Atlas
O. Axis
P. Cervical vertebrae (3–7)
   22. Spinous process
   23. Transverse process
   24. Transverse foramen

Figure 4-3
Axial Skeleton of Cat, Dorsal View

1. Frontal bone
2. Parietal bone
3. Sagittal suture
4. Coronal suture
5. Bregma
6. Atlas
7. Transverse process (wing) of atlas
8. Axis
9. Cervical vertebrae (7)
10. Thoracic vertebrae (13)
11. Ribs
12. Lumbar vertebrae (7)
13. Sacral vertebrae (3)
14. Caudal vertebrae (21–25)
15. Scapula
16. Humerus
17. Ilium
18. Ischium
19. Femur

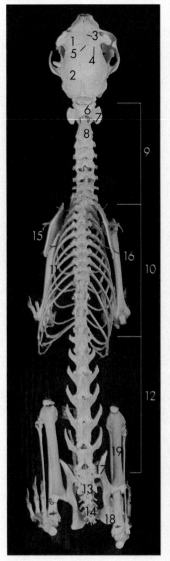

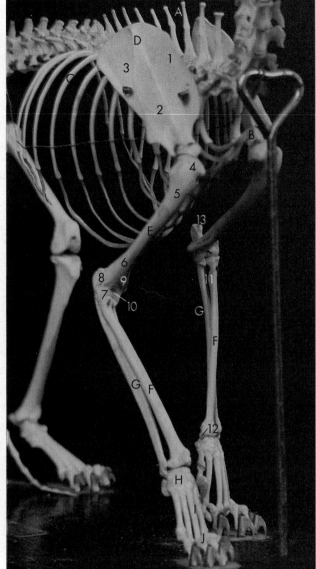

Figure 4-4   Cat Skeleton, Front Right Lateral Aspect

A. Vertebral spinous process
B. Sternum
C. Ribs
D. Scapula
  1. Supraspinous fossa
  2. Acromial spine
  3. Infraspinous fossa
E. Humerus
  4. Proximal head
  5. Deltoid tuberosity
  6. Distal head
  7. Trochlea
  8. Lateral epicondyle
  9. Medial epicondyle
  10. Radial fossa
F. Radius
  11. Radial tuberosity
  12. Styloid process
G. Ulna
  13. Olecranon process
H. Carpal bones
I. Metacarpal bones
J. Phalanges

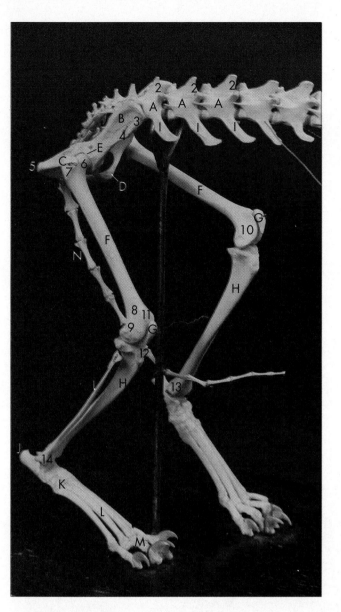

Figure 4-5 Cat Skeleton, Right Lateral Aspect

A. Lumbar vertebra
   1. Transverse process
   2. Spinous process
B. Ilium
   3. Cranial ventral
      iliac spine
   4. Caudal ventral
      iliac spine
C. Ischium
   5. Ischial tuberosity
D. Pubis
E. Acetabulum
F. Femur
   6. Proximal head
   7. Greater trochanter
   8. Distal head

   9. Lateral condyle
   10. Medial condyle
   11. Trochlea
G. Patella
H. Tibia
   12. Tibial tuberosity
   13. Medial malleolus
I. Fibula
   14. Lateral malleolus
J. Calcaneus
K. Tarsal bones
L. Metatarsal bones
M. Phalanges
N. Caudal vertebrae

Figure 4-6
Superficial Anatomy of Cat Head and Neck, Left Lateral View

1. Vibrissal barrels for sensory hairs (whiskers)
2. Tongue
3. Buccinator muscle
4. Diagastric muscle
5. Temporalis muscle
6. Masseter muscle
7. Dorsal buccal branch of facial (VII) nerve
8. Ventral buccal branch of facial (VII) nerve
9. Parotid duct
10. Parotid gland
11. Submaxillary gland
12. Lymph node
13. External jugular vein
14. Transverse jugular vein
15. Anterior facial vein
16. Posterior facial vein
17. Sternohyoid muscle
18. Sternothyroid muscle
19. Cleidomastoid muscle
20. Sternomastoid muscle
21. Clavotrapezius muscle
22. Clavobrachialis muscle
23. Acromiotrapezius muscle

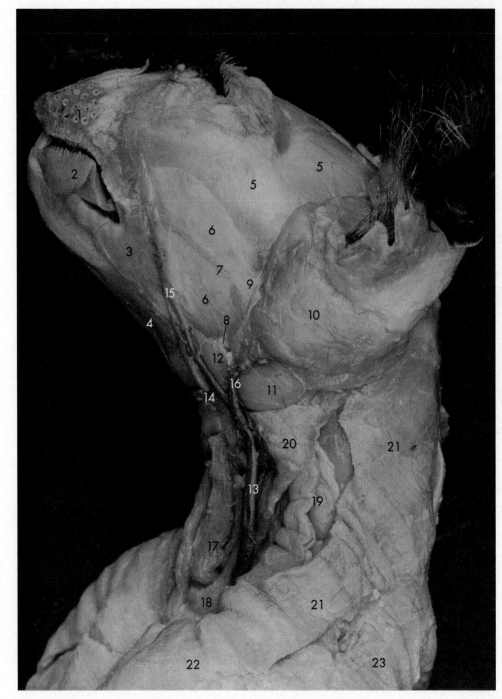

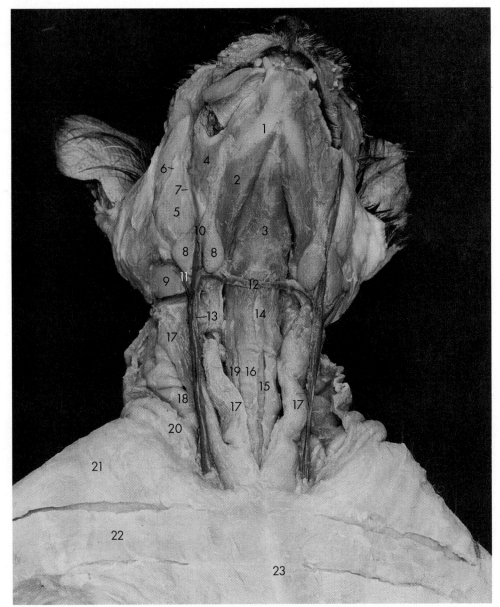

**Figure 4-7
Superficial Anatomy of Cat
Head and Neck, Ventral
Aspect**

1. Body of mandible
2. Digastric muscle
3. Mylohyoid muscle
4. Buccinator muscle
5. Masseter muscle
6. Dorsal branch of facial
   (VII) nerve
7. Ventral branch of facial
   (VII) nerve
8. Lymph node
9. Submaxillary gland
10. Anterior facial vein
11. Posterior facial vein
12. Transverse jugular vein
13. External jugular vein
14. Larynx
15. Trachea
16. Sternohyoid muscle
17. Sternothyroid muscle
    (unavoidably damaged on
    animal's right side during
    vascular perfusion)
18. Cleidomastoid muscle
19. Sternomastoid muscle
20. Clavotrapezius muscle
21. Clavobrachialis muscle
22. Pectoantebrachialis
    muscle
23. Sternum

Figure 4-8
Deep Anatomy of Cat Head and Neck, Left Ventrolateral Aspect

1. Lower canine tooth
2. Upper canine tooth
3. Upper premolar tooth
4. Lower premolar tooth
5. Body of mandible
6. Digastric muscle
7. Mylohyoid muscle
8. Temporalis muscle
9. Masseter muscle
10. Dorsal branch of facial (VII) nerve
11. Ventral branch of facial (VII) nerve
12. Parotid duct
13. Cutaneous branch of facial (VII) nerve
14. Platysma muscle (reflected)
15. Lymph node
16. Sternohyoid muscle
17. Sternothyroid muscle (reflected)
18. Sternomastoid muscle
19. Omohyoid muscle
20. 4th cervical nerve
21. 5th cervical nerve
22. Jugular vein
23. Subclavian vein
24. Musculocutaneous nerve
25. Radial nerve
26. Median nerve
27. Ulnar nerve
28. Thoracic nerve
29. Ventral thoracic nerve (cut)
30. Axillary nerve

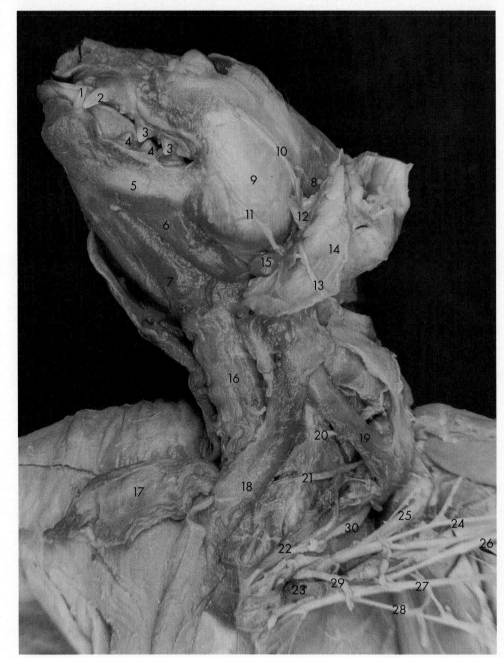

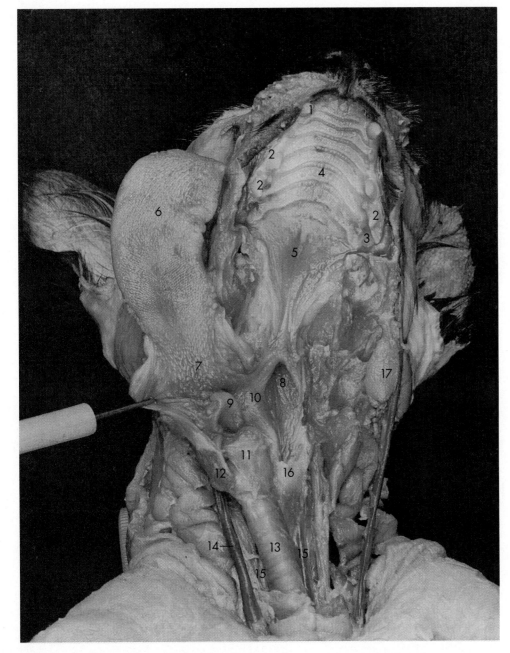

Figure 4-9
Deep Anatomy of Cat Head and Neck, Oral Cavity with Mandible Removed
1. Upper canine tooth
2. Upper premolar tooth
3. Upper molar tooth
4. Hard palate with palatine rugae
5. Soft palate
6. Tongue
7. Foliate papillae
8. Isthmus of Fauces
9. Epiglottis
10. Palatine tonsil
11. Larynx
12. Thyroid gland (reflected)
13. Trachea
14. External jugular vein
15. Carotid artery
16. Esophagus
17. Lymph node

Figure 4-10
Superficial Muscles of the Cat Thoracic Limb, Ventral Aspect

1. Clavobrachialis muscle
2. Pectoantebrachialis muscle
3. Pectoralis major muscle
4. Pectoralis minor muscle
5. Latissimus dorsi muscle
6. Epitrochlearis muscle
7. Flexor carpi ulnaris muscle
8. Palmaris longus muscle
9. Flexor carpi radialis muscle
10. Pronator teres muscle
11. Extensor carpi radialis muscle
12. Brachioradialis muscle (cut)
13. Cephalic vein
14. Antebrachial fascia
15. Ulnar nerve
16. Olecranon process of ulna
17. Flexor retinaculum (transverse carpal ligament)

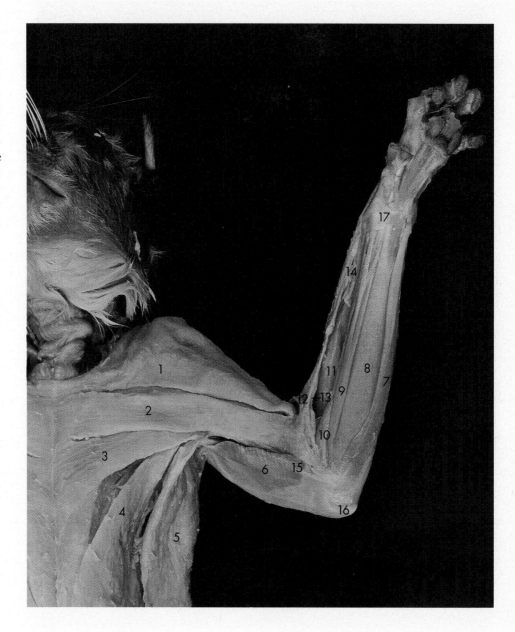

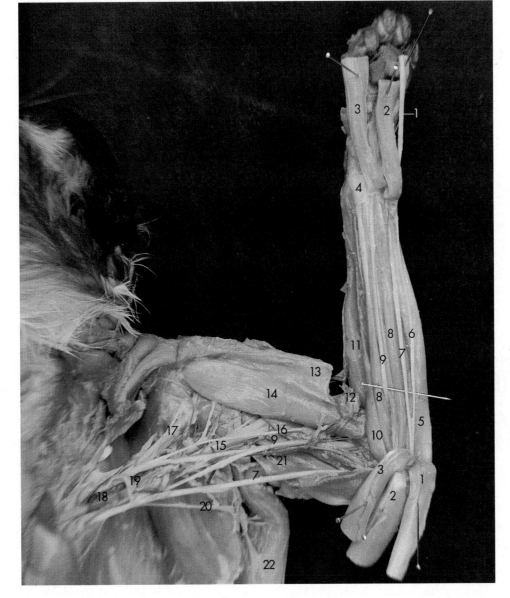

Figure 4-11
Deep Muscles of the Cat Left Thoracic Limb, Ventral Aspect

1. Flexor carpi ulnaris muscle (cut and reflected)
2. Palmaris longus muscle (cut and reflected)
3. Flexor carpi radialis muscle (cut and reflected)
4. Flexor retinaculum
5. Extensor carpi ulnaris
6. Cutaneous branch of ulnar nerve
7. Ulnar nerve
8. Flexor digitorum profundus
9. Median nerve
10. Pronator teres muscle
11. Extensor carpi radialis muscle
12. Brachioradialis muscle (cut)
13. Clavobrachialis muscle (cut and reflected)
14. Biceps brachii muscle
15. Radial nerve
16. Musculocutaneous nerve
17. Axillary nerve
18. Subclavian vein
19. Ventral thoracic nerve (cut)
20. Thoracic nerve
21. Triceps brachii muscle
22. Latissimus dorsi muscle

Figure 4-12   Superficial Muscles of the Cat Left Thoracic Limb, Dorsal Aspect

1. Flexor carpi ulnaris muscle
2. Extensor carpi ulnaris muscle
3. Extensor carpi digitorum lateralis muscle
4. Extensor pollicis brevis muscle
5. Extensor digitorum communis muscle
6. Extensor carpi radialis muscle
7. Brachioradialis muscle
8. Antebrachial fascia
9. Cephalic vein
10. Extensor retinaculum (dorsal carpal ligament)
11. Triceps brachii muscle (lateral head)
12. Triceps brachii muscle (long head)
13. Anconeus muscle
14. Brachialis muscle
15. Clavobrachialis muscle
16. Acromiodeltoid muscle
17. Spinodeltoid muscle
18. Acromiotrapezius muscle
19. Latissimus dorsi muscle
20. Levator scapulae ventralis

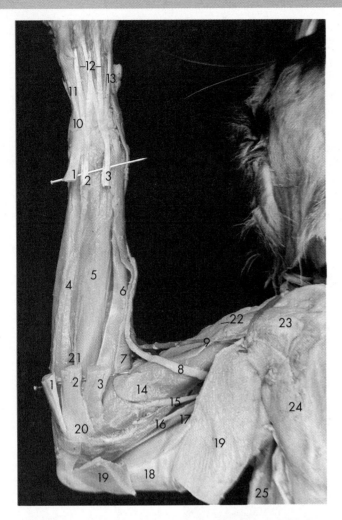

Figure 4-13   Deep Muscles of the Cat Left Thoracic Limb, Dorsal Aspect

1. Extensor carpi ulnaris muscle (cut)
2. Extensor digitorum lateralis muscle (cut)
3. Extensor digitorum communis muscle (cut)
4. Extensor indicus proprius muscle
5. Extensor pollicis brevis muscle
6. Extensor carpi radialis muscle
7. Brachioradialis muscle
8. Radial nerve
9. Cephalic vein
10. Extensor retinaculum (dorsal carpal ligament)
11. Extensor digiti minimi tendon
12. Extensor digitorum tendons
13. Extensor indicis tendon
14. Brachioradialis muscle
15. Median nerve
16. Ulnar nerve
17. Triceps brachii muscle (medial head)
18. Triceps brachii muscle (long head)
19. Triceps brachii muscle (lateral head, cut)
20. Anconeus muscle
21. Posterior interosseous nerve
22. Clavobrachialis muscle
23. Acromiodeltoid muscle
24. Spinodeltoid muscle
25. Latissimus dorsi muscle

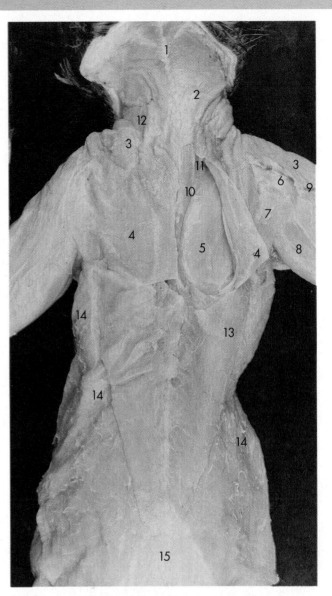

Figure 4-14   Superficial Muscles of the Cat Neck and Back

1. Nuchal ligament
2. Clavotrapezius muscle
3. Clavobrachialis muscle
4. Acromiotrapezius muscle (cut and reflected on right)
5. Supraspinatus muscle
6. Acromiodeltoid muscle
7. Spinodeltoid muscle
8. Triceps brachii muscle (long head)
9. Cephalic vein
10. Rhomboideus minor muscle
11. Rhomboideus capitis muscle (occipito-scapularis muscle)
12. Splenius capitis muscle
13. Spinotrapezius muscle
14. Latissimus dorsi muscle
15. Lumbodorsal fascia

Figure 4-15   Deep Muscles of the Cat Neck and Back

1. Nuchal ligament
2. Clavotrapezius muscle (reflected on left)
3. Acromiotrapezius muscle (cut, removed altogether on left)
4. Supraspinatus muscle
5. Infraspinatus muscle
6. Triceps brachii muscle (long head)
7. Triceps brachii muscle (lateral head)
8. Acromiodeltoid muscle
9. Clavobrachialis muscle
10. Rhomboideus capitis muscle
11. Splenius capitis muscle
12. Rhomboideus minor muscle
13. Rhomboideus major muscle
14. Spinotrapezius muscle
15. Latissimus dorsi muscle (reflected on left, partially removed on right)
16. Multifidus muscle
17. Spinalis muscle
18. Longissimus muscle
19. Iliocostalis muscle
20. Lumbodorsal fascia (largely removed)

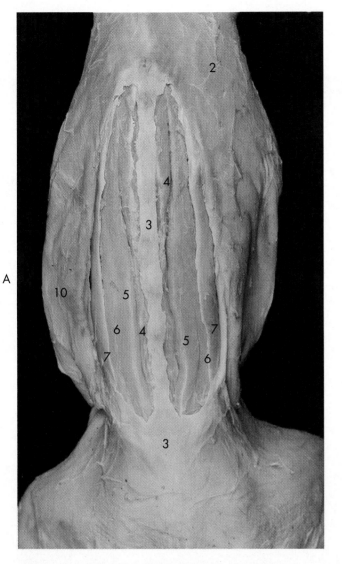

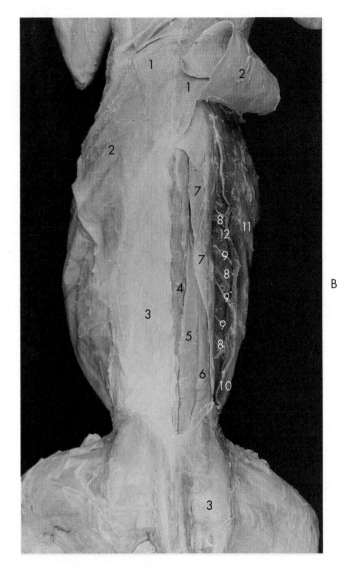

Figure 4-16 Deep Muscles of the Back of the Cat

1. Spinotrapezius muscle
2. Latissimus dorsi muscle (cut and rolled on right)
3. Lumbodorsal fascia
4. Multifidus muscle
5. Spinalis muscle
6. Longissimus muscle
7. Iliocostalis muscle
8. Rib
9. Dorsal ramus of spinal nerve
10. External oblique muscle
11. External intercostal muscles
12. Internal intercostal muscles

Figure 4-17
Superficial Muscles of the Cat Left Hind Limb, Dorsal Aspect

1. Lumbodorsal fascia
2. Sartorius muscle
3. Tensor fascia latae muscle
4. Iliotibial band
5. Gluteus medius muscle
6. Gluteus maximus muscle
7. Caudofemoralis muscle
8. Biceps femoris muscle
9. Semitendinosus muscle
10. Semimembranosus muscle
11. Gastrocnemius muscle
12. Soleus muscle
13. Achilles tendon
14. Calcaneal tuberosity
15. Flexor hallucis longus muscle
16. Peroneus longus muscle
17. Extensor digitorum longus muscle
18. Tibialis anterior muscle

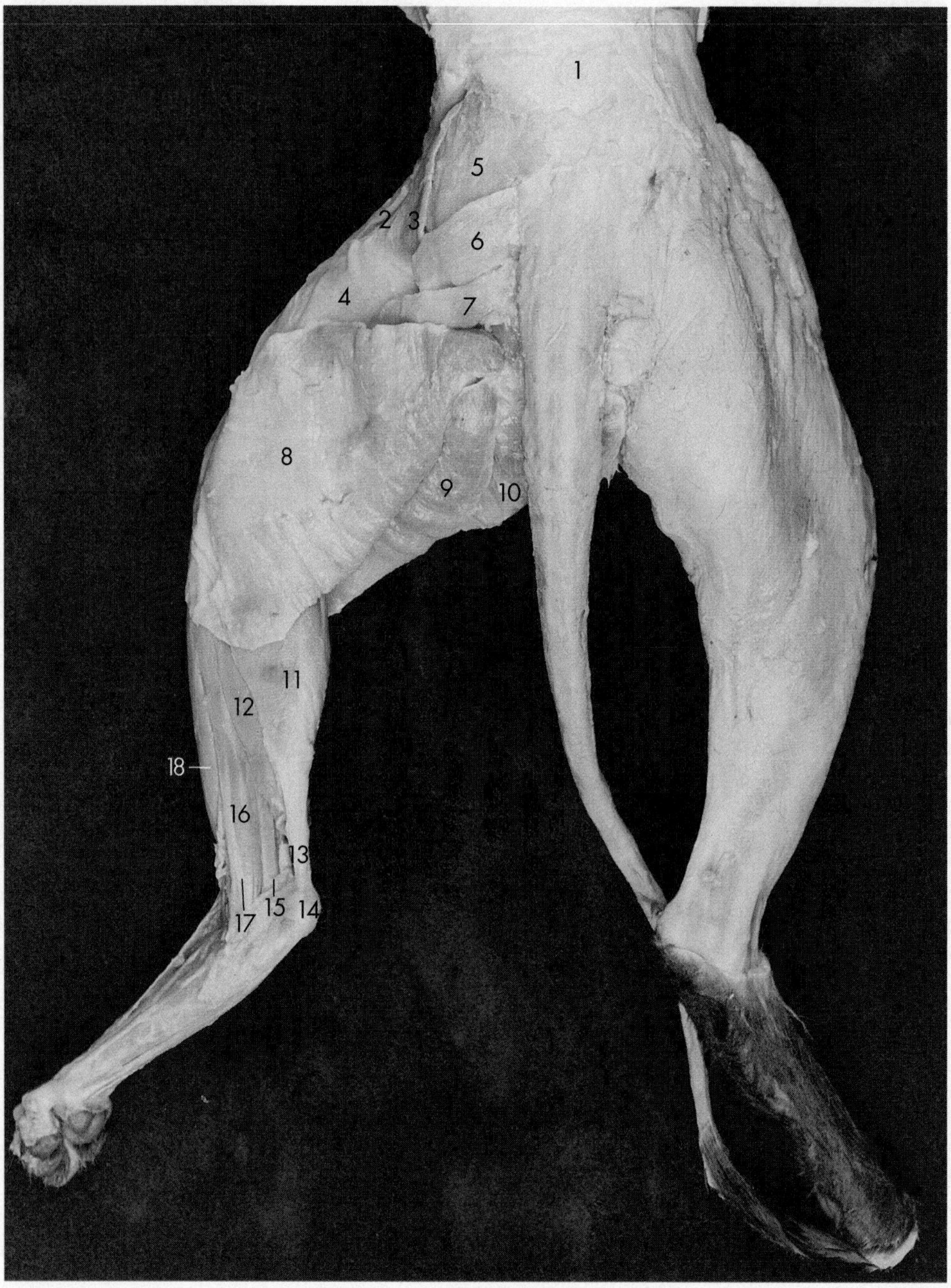

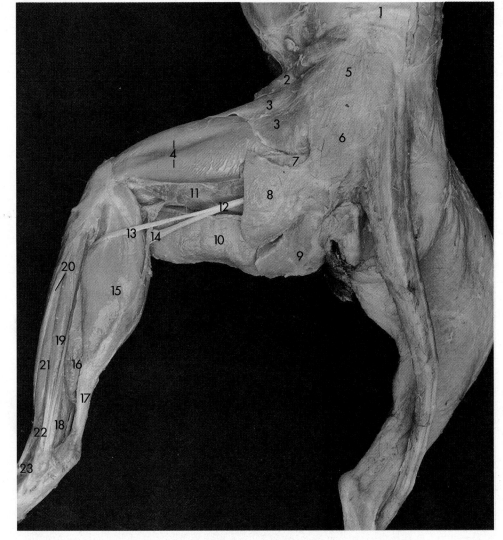

Figure 4-18
Deep Muscles of the Cat Left Hind Limb, Dorsal Aspect

1. Lumbodorsal fascia
2. Sartorius muscle
3. Tensor fascia latae muscle
4. Vastus lateralis muscle
5. Gluteus medius muscle (under fascia)
6. Gluteus maximus muscle (under fascia)
7. Caudofemoralis muscle
8. Biceps femoris muscle (cut)
9. Semitendinosus muscle (cut)
10. Semimembranosus muscle
11. Adductor femoris muscle
12. Sciatic nerve
13. Common peroneal division of sciatic nerve
14. Tibial division of sciatic nerve
15. Gastrocnemius muscle
16. Soleus muscle
17. Achilles tendon
18. Flexor hallucis longus muscle
19. Peroneus longus muscle
20. Tibialis anterior muscle
21. Extensor digitorum longus muscle
22. Proximal extensor retinaculum
23. Distal extensor retinaculum

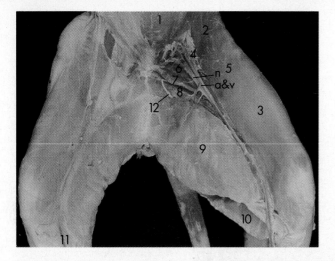
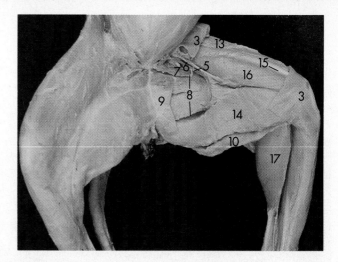

Figure 4-19 Superficial Muscles of the Cat Left Hind Limb, Medial Aspect

1. Rectus abdominis muscle
2. External oblique muscle
3. Sartorius muscle (cut)
4. Iliopsoas muscle (deep to blood vessels)
5. Femoral artery (a), vein (v), and nerve (n)
6. Pectineus muscle (deep to blood vessels)
7. Adductor longus muscle
8. Adductor femoris muscle
9. Gracilis muscle
10. Semitendinosus muscle
11. Greater saphenous vein
12. Branch of obturator nerve
13. Tensor fascia latae muscle
14. Semimembranosus muscle
15. Vastus lateralis muscle
16. ~~Rectus femoris~~ muscle *VASTUS MEDIALIS*
17. Gastrocnemius muscle

Figure 4-20 Deep Muscles of the Cat Left Hind Limb, Medial Aspect

1. Sartorius muscle (cut)
2. Tensor fascia latae muscle
3. Vastus lateralis muscle
4. Vastus medialis muscle
5. Femoral artery (a), vein (v), and nerve (n)
6. Middle caudal femoral artery and vein
7. Pectineus muscle
8. Adductor longus muscle
9. Adductor femoris muscle
10. Gracilis muscle (cut)
11. Semimembranosus muscle
12. Semitendinosus muscle
13. Biceps femoris muscle
14. Gastrocnemius muscle (reflected)
15. Soleus muscle
16. Achilles tendon
17. Posterior tibial nerve
18. Flexor hallucis longus muscle
19. Flexor digitorum longus muscle
20. Tibialis posterior muscle
21. Tibia
22. Tibialis anterior muscle
23. Proximal extensor retinaculum

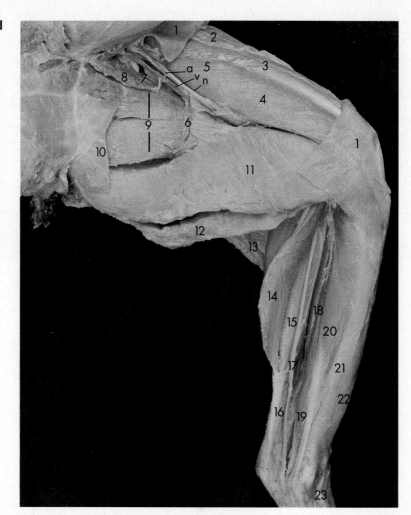

Figure 4-21
Superficial Muscles of the Cat Thorax, Ventral View

1. Clavobrachialis muscle
2. Pectoantebrachialis muscle
3. Pectoralis major muscle
4. Pectoralis minor muscle
5. Xiphihumeralis muscle
6. Epitrochlearis muscle
7. Latissimus dorsi muscle
8. External oblique muscle
9. Rectus abdominis muscle (deep to aponeurosis)
10. Linea alba
11. Inferior angle of scapula
12. Teres major muscle
13. Subscapularis muscle

Figure 4-22
Deep Muscles of the Cat Shoulder and Thorax, Right Ventral View

1. Latissimus dorsi muscle (reflected)
2. Scalenus muscles
   a. Anterior (continuous with transversus costarum)
   b. Medius
   c. Posterior
3. Axillary artery (a) and vein (v)
4. Radial nerve
5. External jugular vein
6. Internal jugular vein
7. Thoracodorsal nerve
8. Long thoracic nerve
9. Thoracoacromial blood vessels
10. Serratus ventralis muscle
11. Teres major muscle
12. Subscapularis muscle
13. Sternum
14. Ventral thoracic nerve

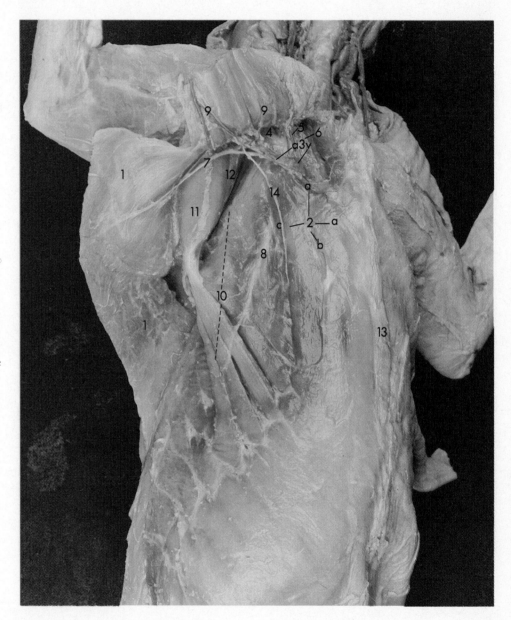

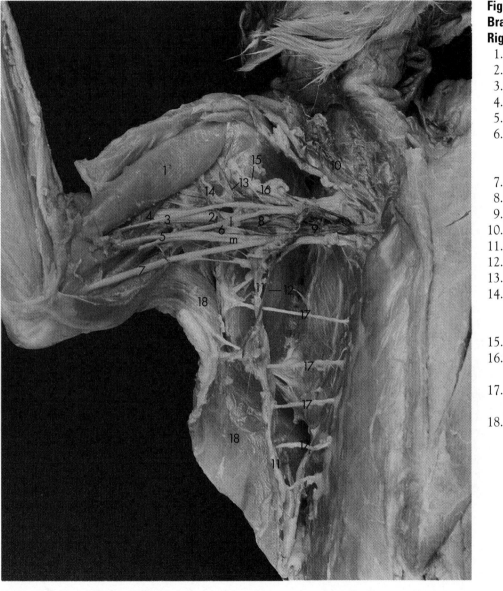

Figure 4-23
Brachial Plexus of the Cat,
Right Ventral Aspect
1. Biceps brachii muscle
2. Radial nerve
3. Musculocutaneous nerve
4. Coracobrachialis muscle
5. Median nerve
6. Lateral (l) and median (m) roots of the median nerve
7. Ulnar nerve
8. Axillary artery
9. Axillary vein
10. External jugular vein
11. Thoracodorsal nerve
12. Thoracodorsal artery
13. Thoracoacromial artery
14. Anterior circumflex humeral artery and axillary nerve
15. Caudal subscapular nerve
16. Proximal subscapular nerve
17. Dorsal rami of thoracic nerves
18. Latissimus dorsi muscle (reflected)

Figure 4-24
Thoracic Cavity of the Cat
1. Heart within pericardium
2. Thymus gland
3. Diaphragm
4. Lung, anterior lobe
5. Lung, middle lobe
6. Lung, posterior lobe
7. Ribs (cut)

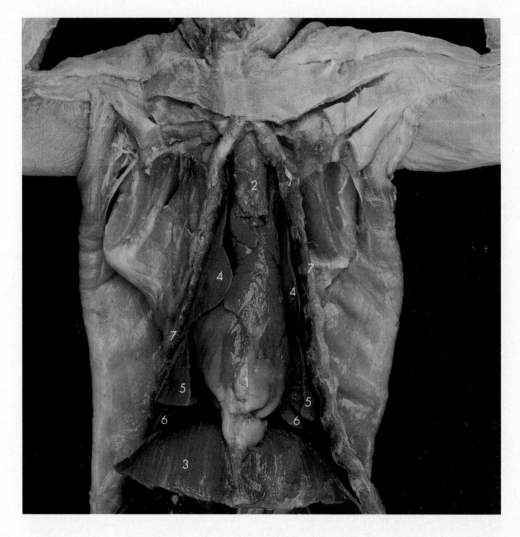

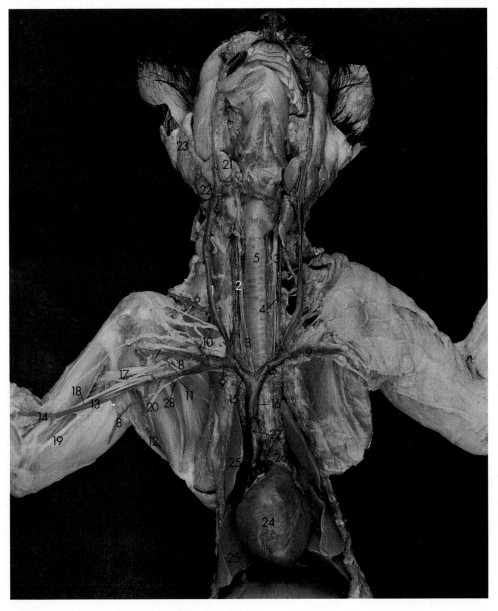

Figure 4-25
Major Veins of the Cat, Neck and Thorax

1. External jugular vein
2. Internal jugular vein
3. Common carotid artery (right)
4. Vagus nerve
5. Trachea
6. Transverse scapular vein
7. Subscapular vein
8. Thoracodorsal vein
9. Subclavian vein
10. Cephalic vein
11. Axillary vein
12. Latissimus dorsi muscle
13. Brachial vein
14. Median cubital vein
15. Innominate (brachiocephalic) vein
16. Anterior vena cava
17. Radial nerve
18. Median nerve
19. Ulnar nerve
20. Thoracodorsal nerve
21. Lymph node
22. Submaxillary gland
23. Parotid gland
24. Heart
25. Lung
26. Thymus gland
27. Anterior thoracic vein (cut) (internal mammary vein)
28. Long thoracic vein (cut)

Figure 4-26
Major Arteries of the Cat,
Neck and Thorax
 1. Common carotid artery
 2. Vagus nerve
 3. Vertebral artery
 4. Transverse scapular artery
 5. Axillary artery
 6. Brachial artery
 7. Thorco-acromial artery
    (a) and nerve (n)
 8. Musculocutaneous nerve
 9. Median nerve
10. Radial nerve
11. Ulnar nerve
12. Trachea
13. Esophagus (displaced to
    animal's left from normal
    position posterior to
    trachea)
14. Phrenic nerve
15. Right subclavian artery
16. Innominate
    (brachiocephalic) artery
17. Left subclavian artery
18. Aortic arch
19. Anterior vena cava (cut)
20. Teres major muscle
21. Subscapularis muscle
22. Biceps brachii muscle
23. Triceps brachii muscle
    (long head)
24. Heart
25. Lung, anterior lobe
26. Lung, middle lobe
27. Lung, mediastinal lobe
28. Lung, posterior lobe
29. Right auricle
30. Diaphragm

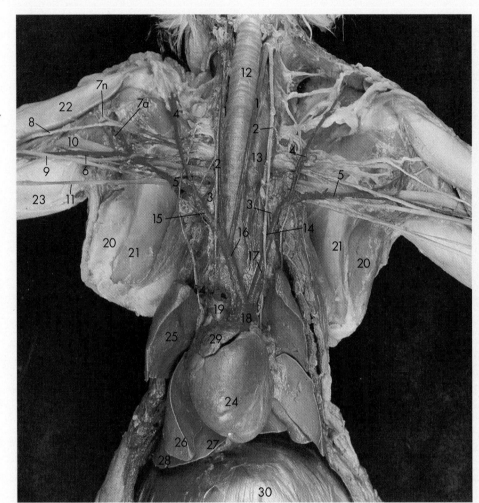

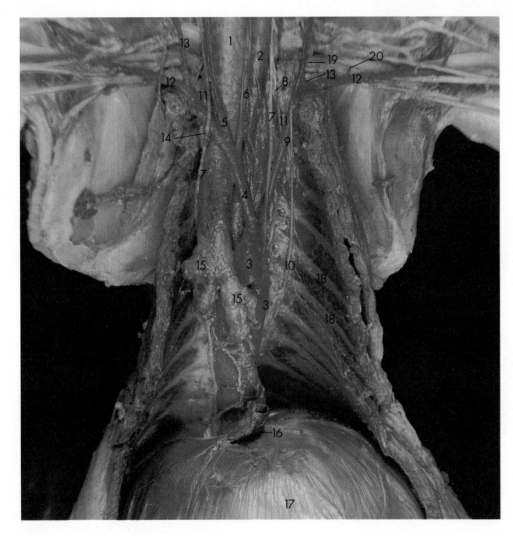

Figure 4-27
Thorax of the Cat, Heart and Lungs Removed
1. Trachea
2. Esophagus
3. Aortic arch
4. Brachiocephalic artery
5. Right common carotid artery
6. Left common carotid artery
7. Vagus nerve
8. Sympathetic trunk
9. Left subclavian artery
10. Phrenic nerve
11. Vertebral artery
12. Subclavian artery
13. Thyrocervical artery
14. Internal mammary artery
15. Right and left primary bronchi
16. Inferior vena cava
17. Diaphragm
18. Rib
19. Transverse scapular artery
20. Subscapular artery (cut)

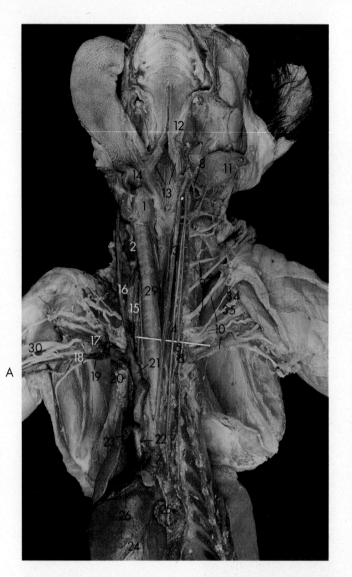

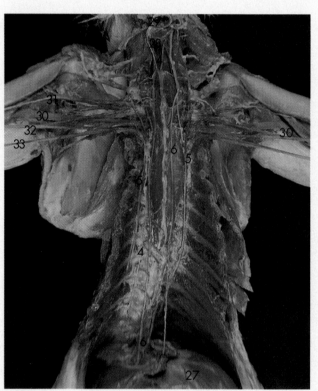

Figure 4-28   Veins, Arteries, and Nerves of the Cat Neck and Thorax
A, Veins Removed on Cat's Left Side, Heart Reflected to Right
B, Arteries, Veins, and Thoracic Viscera Removed

1. Larynx
2. Thyroid gland (reflected)
3. Common carotid artery
4. Vagus nerve
5. Sympathetic trunk (In **A**, two
   pins have been placed along
   sympathetic trunk. Upper pin
   head is just caudal to swelling of
   sympathetic trunk, the superior
   cervical ganglion. Lower
   transverse pin is just proximal to
   similar swelling, the middle
   cervical ganglion.)
6. Phrenic nerve
7. Aorta
8. Spinal accessory nerve (XI)

9. Spinal nerves IV, V, and VI
10. Brachial plexus
11. Lymph node (reflected and
    pinned)
12. Soft palate (cut)
13. Eustachian tubes (hidden behind
    reflected tissue of soft palate)
14. Epiglottis
15. Internal jugular vein
16. External jugular vein
17. Subscapular vein
18. Brachial vein
19. Axillary vein
20. Subclavian vein
21. Innominate (brachiocephalic)
    vein

22. Anterior vena cava
23. Axygous vein (cut)
24. Heart (reflected to cat's right)
25. Right auricle
26. Left auricle
27. Diaphragm
28. Esophagus (cut in **B**)
29. Trachea (cut in **B**)
30. Radial nerve
31. Musculocutaneous nerve
32. Median nerve
33. Ulnar nerve
34. Caudal subscapular nerve
35. Axillary nerve

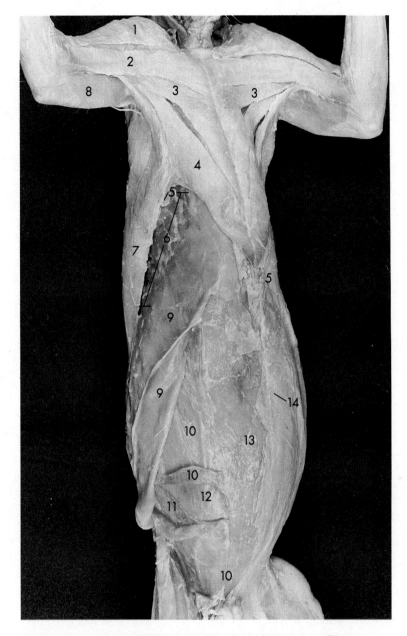

Figure 4-29 Superficial Muscles of the Cat, Abdomen and Thorax

1. Clavobrachialis muscle
2. Pectoantebrachialis muscle
3. Pectoralis major muscle
4. Pectoralis minor muscle
5. Xiphihumeralis muscle (removed on right)
6. Serratus anterior muscle
7. Latissimus dorsi muscle (cut to reveal underlying muscles)
8. Epitrochlearis muscle
9. External oblique muscle (partially reflected)
10. Internal oblique muscle (partially reflected)
11. Transversalis abdominis muscle
12. Peritoneum
13. Rectus abdominis muscle
14. Linea alba

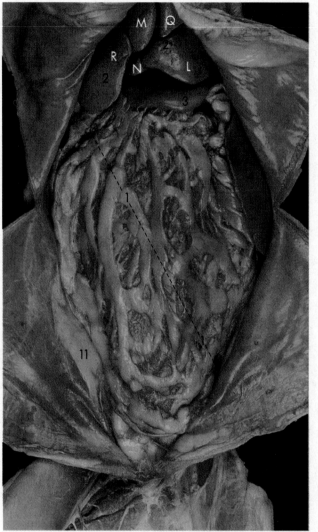
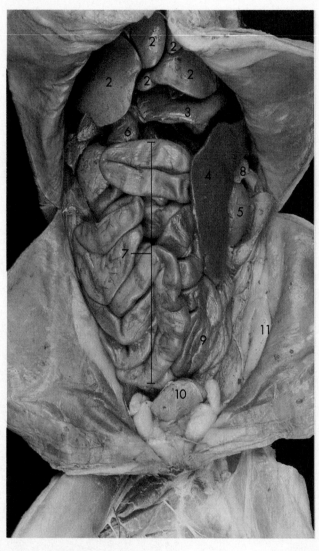

Figure 4-30   Abdominal Viscera of Cat with (A) Greater Omentum Intact and (B) Greater Omentum Removed

1. Greater omentum
2. Lobes of the liver
   R. Right lateral lobe
   M. Right medial lobe
   Q. Quadrate lobe
   N. Left medial lobe
   L. Left lateral lobe
3. Stomach (greater curvature)
4. Spleen

5. Kidney
6. Small intestine (duodenum)
7. Small intestine (jejunem and ileum)
8. Pancreas
9. Large intestine (descending colon)
10. Bladder
11. Abdominal fat

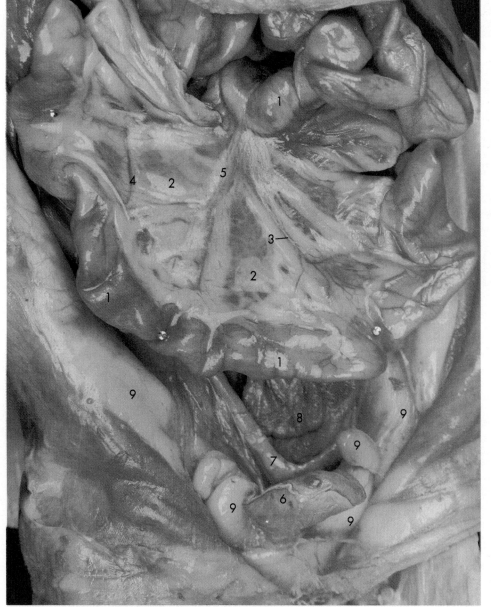

Figure 4-31
Abdominal Viscera of the Cat, Mesentery
1. Small intestine
2. Mesentery
3. Mesenteric artery
4. Mesenteric vein
5. Lymph vessel
6. Bladder
7. Uterus
8. Rectum
9. Abdominal fat

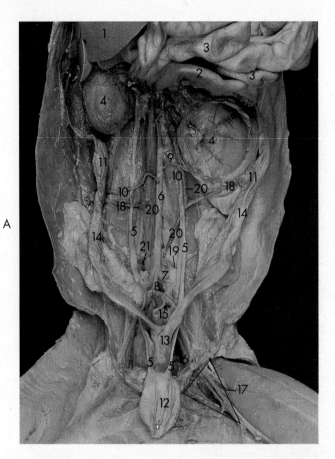

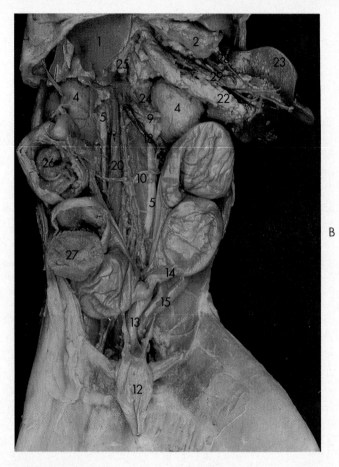

Figure 4-32   Urogenital System of the Female Cat (A) Nonpregnant and (B) Pregnant

1. Liver
2. Stomach (pylorus)
3. Small intestine
4. Kidney
5. Ureter
6. Abdominal aorta
7. Common iliac artery
8. Caudal (median sacral) artery
9. Renal artery
10. Ovarian artery
11. Ovary

12. Bladder (reflected and pinned)
13. Uterus
14. Uterine horn (in **B**, left horn contains two fetuses; right horn, three fetuses)
15. Rectum (cut in **A**)
16. External iliac artery and vein
17. Femoral triangle (containing femoral nerve, artery, and vein)
18. Ovarian vein
19. Iliolumbar vein

20. Abdominal vena cava (split into two parallel vessels in **A**)
21. Iliolumbar artery
22. Pancreas
23. Spleen
24. Adrenal gland
25. Hepatic portal vein (cut)
26. Fetus
27. Placenta
28. Left gastroepiploic vein
29. Right gastroepiploic vein

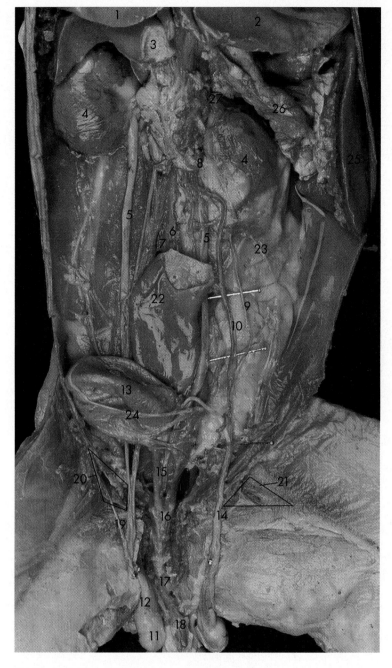

Figure 4-33 Urogenital System of the Male Cat

1. Liver
2. Stomach
3. Small intestine (duodenum, cut)
4. Kidney
5. Ureter
6. Abdominal aorta
7. Abdominal vena cava
8. Renal artery
9. Internal spermatic artery
10. Spermatic vein
11. Testis
12. Epididymis
13. Bladder (reflected)
14. Vas deferens in spermatic cord

15. Urethra
16. Prostate gland
17. Bulbourethral (Cowper's) gland
18. Penis
19. Ligament of cremaster muscle
20. External inguinal ring
21. Femoral triangle
22. Rectum (cut)
23. Lumbar nerve (medial branch)
24. Umbilical (allantoic) artery
25. Spleen
26. Pancreas
27. Adrenal gland

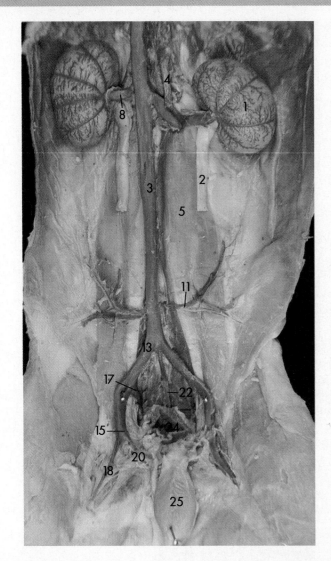
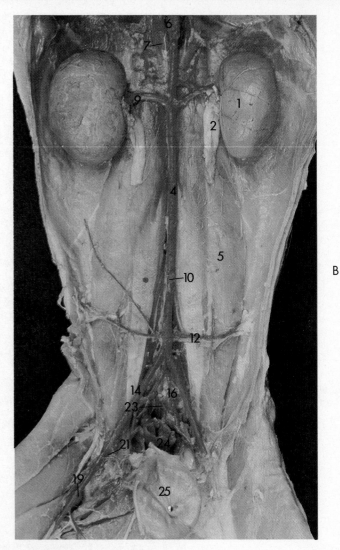

Figure 4-34   Major (A) Veins and (B) Arteries of the Cat Abdominopelvic Wall

1. Kidney
2. Ureter (cut and largely removed)
3. Abdominal vena cava (removed in **B**)
4. Abdominal aorta (cut and removed in **A**)
5. Psoas muscle
6. Celiac artery (cut and removed)
7. Superior mesenteric artery (cut and removed)

8. Renal vein
9. Renal artery
10. Inferior mesenteric artery (cut and removed)
11. Iliolumbar vein
12. Iliolumbar artery
13. Common iliac vein
14. External iliac artery (no common iliac artery in cats)
15. External iliac vein

16. Interior iliac (hypogastric) artery
17. Internal iliac (hypogastric) vein
18. Femoral vein
19. Femoral artery
20. Deep femoral vein
21. Deep femoral artery
22. Caudal vein
23. Median sacral (caudal) artery
24. Rectum (cut)
25. Bladder (reflected and pinned)

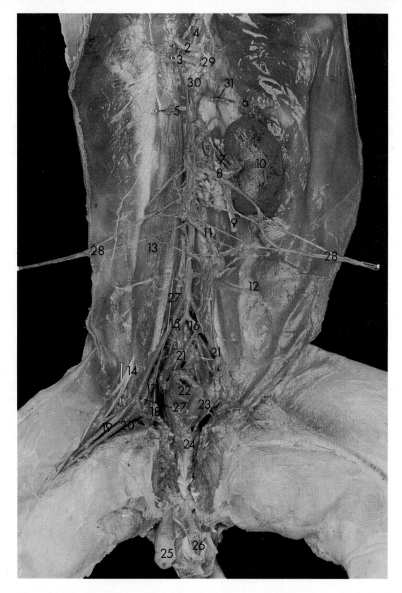

Figure 4-35 Nerves and Vessels of the Posterior Abdominopelvic Wall of the Cat

1. Abdominal aorta
2. Celiac artery (pinned)
3. Superior mesenteric artery (pinned)
4. Crus of diaphragm
5. Right adrenolumbar (phrenicoabdominal) artery (cut)
6. Adrenal gland
7. Renal vein
8. Renal artery
9. Ureter (cut)
10. Kidney
11. Inferior mesenteric artery
12. Iliolumbar artery
13. Psoas muscle
14. Femoral nerve
15. External iliac artery
16. Internal iliac artery
17. External iliac vein
18. Deep femoral artery and vein
19. Femoral artery
20. Femoral vein
21. Spermatic artery
22. Rectum (cut)
23. Urethra (cut and bladder removed)
24. Prostate gland
25. Testis
26. Penis
27. Genitofemoral nerve
28. Distribution of sympathetic trunk (pinned out bilaterally)
29. Celiac ganglion
30. Superior mesenteric ganglion
31. Left adrenolumbar artery and vein

Figure 4-36
Superficial Muscles of the Fetal Pig, Left Lateral View

1. Clavotrapezius muscle
2. Clavobrachialis muscle
3. Acromiodeltoid muscle
4. Spinodeltoid muscle
5. Triceps brachii muscle
6. Spinotrapezius muscle (cut)
7. Latissimus dorsi muscle
8. External oblique muscle (cut)
9. Serratus anterior muscle
10. Internal oblique muscle
11. Tensor fascia latae muscle (split)
12. Vastus lateral muscle (under pin)
13. Gluteus medius muscle
14. Gluteus maximus muscle
15. Biceps femoris muscle
16. Semitendinosus muscle
17. Semimembranosus muscle
18. Testis
19. Umbilical cord

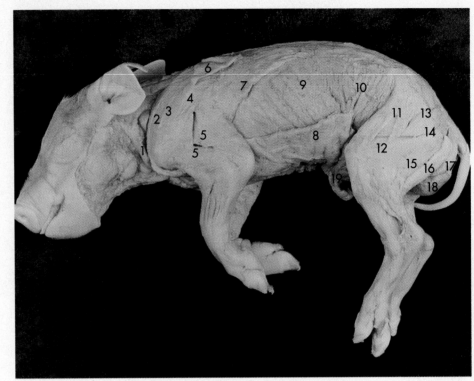

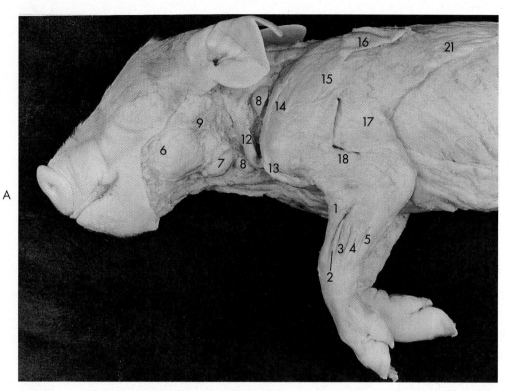

A

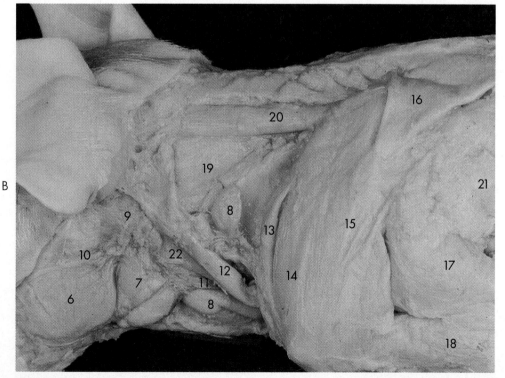

B

Figure 4-37
Superficial Structures of the Neck, Shoulder, and Thoracic Limb of the Fetal Pig, Left Lateral View

1. Brachioradialis muscle
2. Extensor carpi radialis muscle
3. Extensor digitorum communis muscle
4. Extensor digitorum lateralis muscle
5. Extensor carpi ulnaris muscle
6. Masseter muscle
7. Submaxillary gland
8. Lymph node
9. Parotid gland
10. Salivary duct
11. External jugular vein
12. Clavotrapezius muscle
13. Clavobrachialis muscle
14. Acromiodeltoid muscle
15. Spinodeltoid muscle
16. Spinotrapezius muscle (cut)
17. Triceps brachii muscle (long head)
18. Triceps brachii muscle (lateral head)
19. Splenius capitis muscle
20. Rhomboideus capitis muscle
21. Latissimus dorsi muscle
22. Sternomastoid muscle

Figure 4-38
Superficial Muscles of the
Hind Limb of the Fetal Pig,
Left Lateral View

1. Lumbodorsal fascia
2. External oblique muscle (reflected)
3. Internal oblique muscle
4. Tensor fascia latae muscle (split)
5. Vastus lateralis muscle (under pin)
6. Gluteus medius muscle
7. Gluteus maximus muscle
8. Biceps femoris muscle
9. Semitendinosus muscle
10. Semimembranosus muscle
11. Testis
12. Gastrocnemius muscle
13. Soleus muscle
14. Achilles tendon
15. Flexor hallucis longus muscle
16. Tibialis anterior muscle

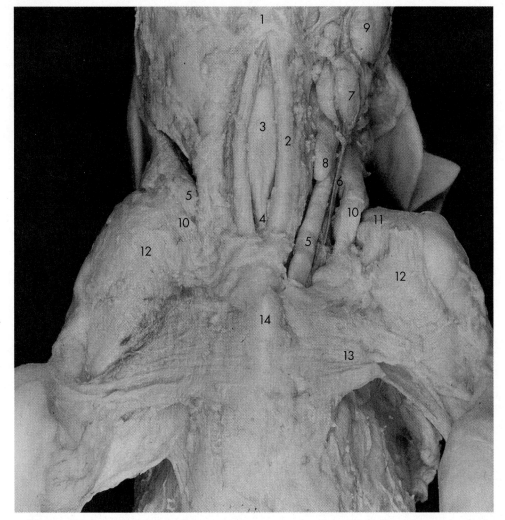

Figure 4-39
Superficial Anatomy of the
Fetal Pig Neck and
Shoulders, Ventral Aspect
  1. Mylohyoid muscle
  2. Sternohyoid muscle
  3. Larynx
  4. Trachea
  5. Sternomastoid muscle
  6. External jugular vein
  7. Lymph node
  8. Submaxillary gland
  9. Masseter muscle
10. Clavotrapezius muscle
11. Acromiodeltoid muscle
12. Clavobrachialis muscle
13. Pectoralis major muscle
14. Sternum

Figure 4-40
Deep Anatomy of the Fetal
Pig, Neck and Thorax

1. Larynx
2. Trachea
3. Thyroid gland
4. Common carotid artery
5. Vagus nerve
6. Internal jugular vein
7. External jugular vein
   (pinned bilaterally)
8. Cephalic vein
9. Subclavian vein
10. Superior vena cava
11. Internal mammary vein
    (cut, laid on lung tissue)
12. Right auricle
13. Left auricle
14. Heart
15. Lung
16. Diaphragm

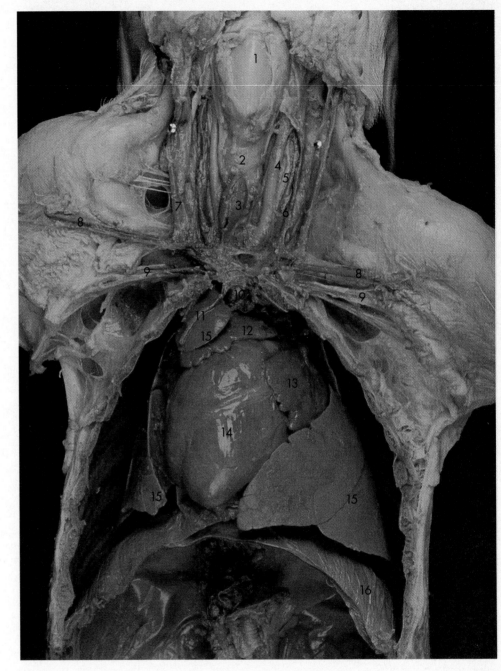

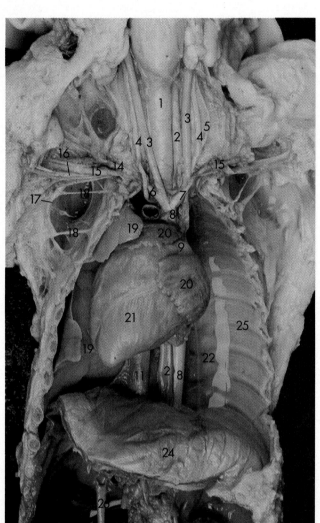

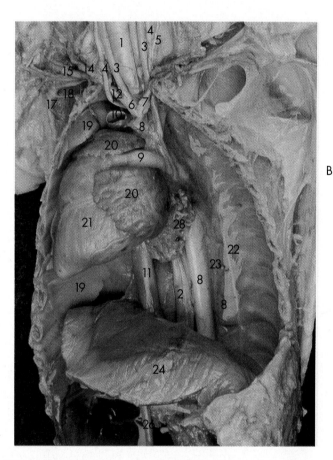

A   B

Figure 4-41   Arteries of the Neck and Thorax of the Fetal Pig, Left Lung Removed
A, Heart in Normal Position
B, Heart Reflected to Right

1. Trachea
2. Esophagus
3. Common carotid artery
4. Vagus nerve
5. Sympathetic trunk
6. Right innominate
   (brachiocephalic) artery
7. Left innominate
   (brachiocephalic) artery
8. Aorta
9. Ductus arteriosus

10. Superior vena cava (cut)
11. Inferior vena cava
12. Subclavian artery
13. Vertebral artery
14. Transverse scapular artery
15. Axillary artery
16. Radial nerve
17. Thoracodorsal nerve
18. Dorsal rami of thoracic nerves
19. Lung
20. Right and left auricles

21. Heart
22. Continuation of sympathetic
    trunk
23. Azygous vein
24. Diaphragm
25. Rib with costal artery and vein
26. Ductus venosus
27. Kidney
28. Hilum of left lung (with bronchi
    and blood vessels cut)

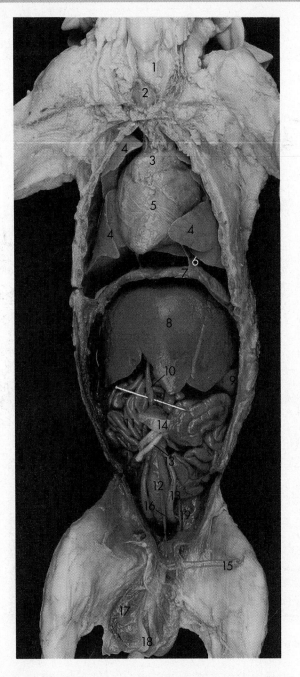

Figure 4-42   Thoracic and Abdominal Viscera of the Fetal Pig

1. Trachea
2. Thyroid gland
3. Thymus
4. Lung
5. Heart in pericardium
6. Mediastinal membrane
7. Diaphragm
8. Liver
9. Spleen
10. Ductus venosus (on pin)
11. Small intestine
12. Bladder
13. Umbilical arteries
14. Umbilical vein
15. Penis
16. Urethra
17. Testis
18. Epididymis
19. Spermatic cord (contains spermatic artery and vas deferens, which curves to pass behind base of bladder)

Figure 4-43   Abdominopelvic Cavity of the Fetal Pig, Male, Digestive Viscera Removed

1. Abdominal aorta
2. Abdominal vena cava
3. Renal artery
4. Renal vein
5. Kidney
6. Ureter
7. Spermatic artery
8. a. Vas deferens
   b. Vas deferens (in spermatic cord)
9. Testis
10. Epididymis
11. Penis
12. Rectum (cut)
13. Umbilical arteries
14. Umbilical vein
15. Ductus venosus (only remnants of liver remain)
16. Bladder
17. Urethra
18. Prostate gland
19. Diaphragm
20. Heart
21. Lung
22. Pylorus of stomach (pin in antrum)

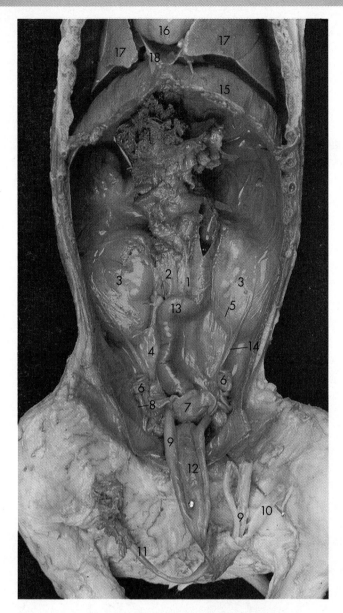

Figure 4-44   Abdominopelvic Cavity of the Fetal Pig, Female, Digestive Viscera Removed

1. Abdominal aorta
2. Abdominal vena cava
3. Kidney (behind intact peritoneum)
4. Ureter (behind intact peritoneum)
5. Ovarian artery
6. Ovary
7. Uterus
8. Uterine horn
9. Umbilical arteries
10. Umbilical vein (lying on pin)
11. Ductus venosus
12. Bladder (reflected and pinned)
13. Sigmoid colon
14. Suspensory ligament of ovary
15. Diaphragm
16. Heart
17. Lung
18. Mediastinal membrane

Figure 4-45   Deep Anatomy of the Abdominopelvic Cavity of the Fetal Pig, Abdominal Viscera Removed, Female

1. Abdominal aorta
2. Abdominal vena cava
3. Renal vein
4. Ureter
5. Kidney (left kidney behind peritoneum)
6. Adrenal gland
7. Suspensory ligament of ovary
8. External iliac artery
9. Internal iliac artery
10. Median sacral (caudal) artery
11. Rectum (cut)
12. Ovary
13. Uterine horn
14. Bladder (reflected)
15. Urethra
16. Umbilical arteries
17. Umbilical vein
18. Ductus venosus
19. Posterior (inferior) mesenteric artery
20. Colic artery (pinned to kidney for clarity due to missing colon)
21. Superior hemorrhoidal artery
22. Remnant of small intestine (duodenum)
23. Diaphragm

Figure 4-46
General Anatomy of the Male Rat, Abdominal Cavity Exposed, Ventral View

1. Thorax
2. Abdomen
3. External oblique muscle (reflected and pinned)
4. Internal oblique muscle (lying on pin)
5. Transversus abdominis
6. Rectus abdominis
7. Peritoneum
8. Inferior epigastric artery
9. Liver
10. Spleen
11. Kidney
12. Rectum
13. Abdominal fat (small intestine not visible in this photograph)
14. Testis
15. Penis
16. Sternum (xiphoid process)

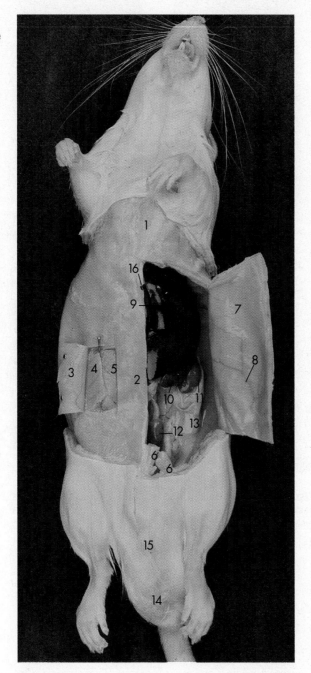

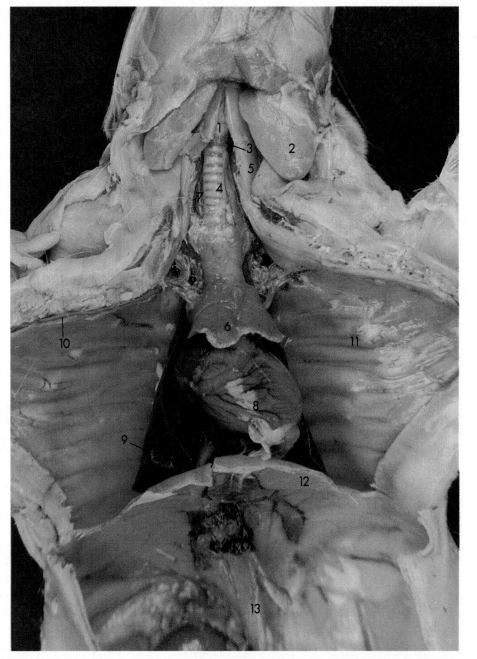

Figure 4-47
Deep Anatomy of the Rat,
Neck and Thorax
1. Larynx
2. Salivary gland
3. Thyroid gland
4. Trachea
5. Sternohyoid muscle
   (unavoidably damaged on
   animal's right side during
   vascular perfusion)
6. Thymus
7. Common carotid artery
8. Heart
9. Lung
10. Internal mammary vein
11. Rib and intercostal artery
    and vein
12. Diaphragm
13. Crus of diaphragm

Figure 4-48
Abdominopelvic Cavity of the Male Rat

1. Sternum (xiphoid process)
2. Stomach
3. Liver
4. Small intestine (duodenum)
5. Pancreas
6. Spleen
7. Kidney
8. Small intestine (jejunem and ileum)
9. Large intestine (cecum)
10. Rectum
11. Abdominal fat
12. Bladder
13. Rectus abdominis muscle (cut)
14. Testis in scrotum
15. Epididymis
16. Penis
17. Seminal vesicle
18. Cremasteric fascia

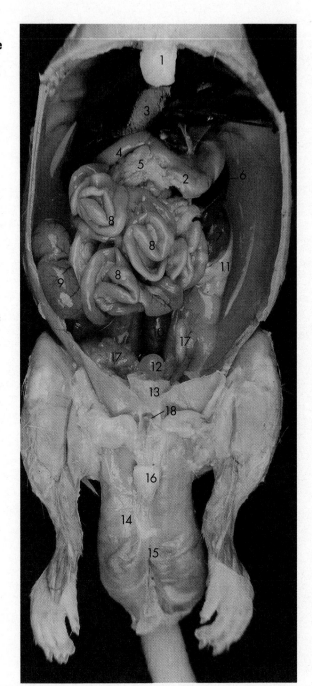

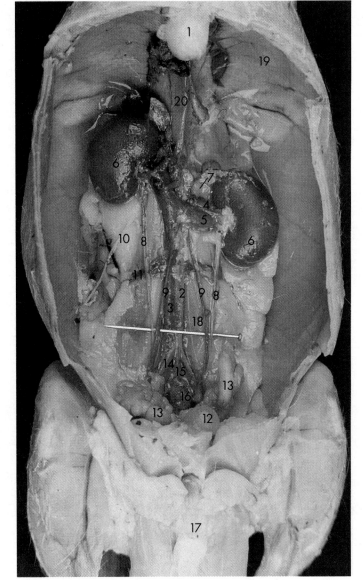

Figure 4-49
Abdominopelvic Cavity of the Rat, Male, Digestive Viscera Removed
1. Sternum (xiphoid process)
2. Abdominal aorta
3. Abdominal vena cava
4. Renal artery
5. Renal vein
6. Kidney
7. Adrenal gland
8. Ureter (lying on pin)
9. Spermatic artery (lying on pin)
10. Lumbar nerve (medial branch, extended for clarity)
11. Iliolumbar artery and vein
12. Bladder
13. Seminal vesicle
14. Common iliac artery
15. Median sacral (caudal) artery
16. Rectum (cut)
17. Penis
18. Psoas muscle
19. Diaphragm
20. Crus of diaphragm

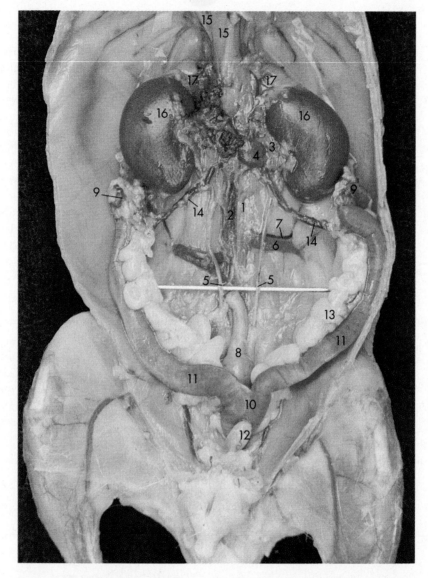

Figure 4-50   Abdominopelvic Cavity of the Rat, Female, Digestive Viscera Removed

1. Abdominal aorta
2. Abdominal vena cava
3. Renal artery
4. Renal vein
5. Ureter (lying on pin)
6. Iliolumbar artery
7. Iliolumbar vein
8. Rectum (cut)
9. Ovary
10. Uterus
11. Uterine horn
12. Bladder
13. Abdominal fat
14. Ovarian artery and vein
15. Crus of diaphragm
16. Kidney
17. Adrenal gland

CHAPTER 5
REFERENCE TABLES

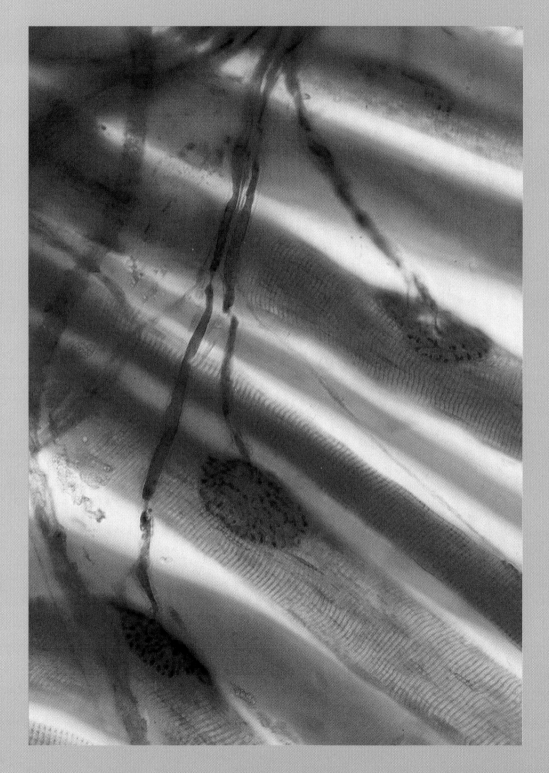

Innervation of Skeletal Muscle: Motor Endplate

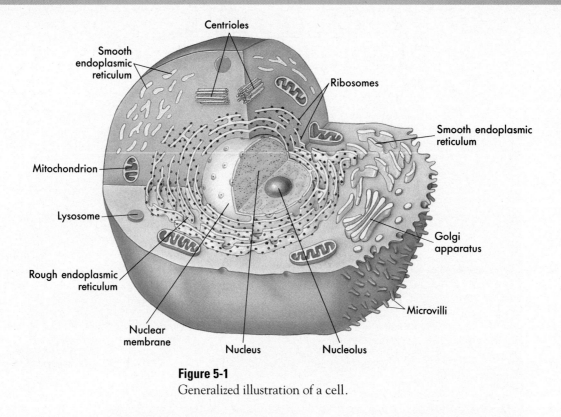

Figure 5-1
Generalized illustration of a cell.

TABLE 5-1 Some major cell structures and their functions (see Figure 5-1)

Cell Structure	Functions
MEMBRANOUS	
Plasma membrane	Serves as the boundary of the cell, maintaining its integrity; protein molecules on outer surface of plasma membrane perform various functions; for example, they serve as markers that identify cells of each individual, as receptor molecules for certain hormones and other molecules, and as transport mechanisms
Endoplasmic reticulum (ER)	Ribosomes attached to rough ER synthesize proteins that leave cells via the Golgi complex; smooth ER synthesizes lipids incorporated in cell membranes, steroid hormones, and certain carbohydrates used to form glycoproteins
Golgi apparatus	Synthesizes carbohydrate, combines it with protein, and packages the product as globules of glycoprotein
Lysosomes	A cell's "digestive system"
Peroxisomes	Contain enzymes that detoxify harmful substances
Mitochondria	Catabolism; ATP synthesis; a cell's "power plants"
Nucleus	Dictates protein synthesis, thereby playing essential role in other cell activities, namely, cell transport, metabolism, and growth
NONMEMBRANOUS	
Ribosomes	Synthesize proteins; a cell's "protein factories"
Cytoskeleton	Acts as a framework to support the cell and its organelles; functions in cell movement; forms cell extensions (microvilli, cilia, flagella)
Cilia and Flagella	Hairlike cell extensions that serve to move substances over the cell surface (cilia) or propel sperm cells (flagella)
Nucleolus	Plays an essential role in the formation of ribosomes

Figure 5-1 and all tables are from Thibodeau GA, Patton KT: *Anatomy and Physiology*, ed 2, St Louis, 1993, Mosby.

TABLE 5-2 Some important transport processes

Process	Type	Description		Examples
Simple Diffusion	Passive	Movement of particles through the phospholipid bilayer or through channels from an area of high concentration to an area of low concentration—that is, down the concentration gradient		Movement of carbon dioxide out of all cells; movement of sodium ions into nerve cells as they conduct an impulse
Dialysis	Passive	Diffusion of small solute particles, but not larger solute particles, through a selectively permeable membrane; results in separation of large and small solutes		During procedure called *peritoneal dialysis,* small solutes diffuse from blood vessels but blood proteins do not (thus removing only small solutes from the blood)
Osmosis	Passive	Diffusion of water through a selectively permeable membrane in the presence of at least one impermeant solute		Diffusion of water molecules into and out of cells to correct imbalances in water concentration
Facilitated Diffusion	Passive	Diffusion of particles through a membrane by means of carrier molecules; also called *carrier-mediated passive transport*		Movement of glucose molecules into most cells
Active Transport	Active	Movement of solute particles from an area of low concentration to an area of high concentration (up the concentration gradient) by means of a carrier molecule		In muscle cells, pumping of nearly all calcium ions to special compartments—or out of the cell
Phagocytosis	Active	Movement of cells or other large particles into a cell by trapping it in a section of plasma membrane that pinches off to form an intracellular vesicle; type of *endocytosis*		Trapping of bacterial cells by phagocytic white blood cells
Pinocytosis	Active	Movement of fluid and dissolved molecules into a cell by trapping them in a section of plasma membrane that pinches off to form an intracellular vesicle; type of *endocytosis*		Trapping of large protein molecules by some body cells
Exocytosis	Active	Movement of proteins or other cell products out of the cell by fusing a secretory vesicle with the plasma membrane		Secretion of the hormone, prolactin, by pituitary cells

TABLE 5-3 Bones of skeleton (206 total)* (see Figure 2-1)

AXIAL SKELETON (80 bones total)		AXIAL SKELETON—cont'd	
Part of Body	**Name of Bone**	**Part of Body**	**Name of Bone**
Skull (28 bones total)		Sternum and ribs (25 bones total)	Sternum (1)
Cranium (8 bones)	Frontal (1)		True ribs (14)
	Parietal (2)		False ribs (10)
	Temporal (2)	**APPENDICULAR SKELETON (126 bones total)**	
	Occipital (1)	**Part of Body**	**Name of Bone**
	Sphenoid (1)	Upper extremities	
	Ethmoid (1)	(including shoulder girdle) (64 bones total)	Clavicle (2)
Face (14 bones)	Nasal (2)		Scapula (2)
	Maxillary (2)		Humerus (2)
	Zygomatic (malar)((2)		Radius (2)
	Mandible (1)		Ulna (2)
	Lacrimal (2)		Carpals (16)
	Palatine (2)		Metacarpals (10)
	Inferior conchae (turbinates) (2)		Phalanges (28)
	Vomer (1)	Lower extremities (including hip girdle) (62 bones total)	Coxal bones (2)
Ear bones (6 bones)	Malleus (hammer) (2)		Femur (2)
	Incus (anvil) (2)		Patella (2)
	Stapes (stirrup) (2)		Tibia (2)
Hyoid bone (1)			Fibula (2)
Spinal column (26 bones total)	Cervical vertebrae (7)		Tarsals (14)
	Thoracic vertebrae (12)		Metatarsals (10)
	Lumbar vertebrae (5)		Phalanges (28)
	Sacrum (1)		
	Coccyx (1)		

*An inconstant number of small, flat, round bones known as **sesamoid bones** (because of their resemblance to sesame seeds) are found in various tendons in which considerable pressure develops. Because the number of these bones varies greatly between individuals, only two of them, the patellae, have been counted among the 206 bones of the body. Generally, two of them can be found in each thumb (in flexor tendon near metacarpophalangeal and interphalangeal joints) and great toe plus several others in the upper and lower extremities. **Wormian bones,** the small islets of bone frequently found in some of the cranial sutures, have not been counted in this list of 206 bones because of their variable occurrence.

TABLE 5-4 Comparison of male and female skeletons

Portion of Skeleton	Male	Female
GENERAL FORM	Bones heavier and thicker Muscle attachment sites more massive Joint surfaces relatively large	Bones lighter and thinner Muscle attachment sites less distinct Joint surfaces relatively small
SKULL	Forehead shorter vertically Mandible and maxillae relatively larger Facial area more pronounced Processes more prominent	Forehead more elongated vertically Mandible and maxillae relatively smaller Facial area rounder, with less pronounced features Processes less pronounced
PELVIS		
Pelvic cavity	Narrower in all dimensions Deeper Pelvic outlet relatively small	Wider in all dimensions Shorter and Roomier Pelvic outlet relatively large
Sacrum	Long, narrow, with smooth concavity (sacral curvature); sacral promontory more pronounced	Short, wide, flat concavity more pronounced in a posterior direction; sacral promontory less pronounced
Coccyx	Less movable	More movable and follows posterior direction of sacral curvature
Pubic arch	Less than a 90° angle	Greater than a 90° angle
Symphysis pubis	Relatively deep	Relatively shallow
Ischial spine, ischial tuberosity, and anterior superior iliac spine	Turned more inward	Turned more outward and further apart
Greater sciatic notch	Narrow	Wide

TABLE 5-5 Major events of muscle contraction and relaxation

EXCITATION AND CONTRACTION

1. A nerve impulse reaches the end of a motor neuron, triggering the release of the neurotransmitter *acetylcholine.*
2. Acetylcholine diffuses rapidly across the gap of the neuromuscular junction and binds to acetylcholine receptors on the motor endplate of the muscle fiber.
3. Stimulation of acetylcholine receptors initiates an impulse that travels along the sarcolemma, through the T tubules, to sacs of the sarcoplasmic reticulum (SR).
4. Ca^{++} is released from the SR into the sarcoplasm, where it binds to troponin molecules in the thin myofilaments.
5. Tropomyosin molecules in the thin myofilaments shift, exposing actin's active sites.
6. Energized myosin cross bridges of the thick myofilaments bind to actin and use their energy to pull the thin myofilaments toward the center of each sarcomere. This cycle repeats itself many times per second, as long as ATP is available.
7. As the thin filaments slide past the thick myofilaments, the entire muscle fiber shortens.

RELAXATION

1. After the impulse is over, the SR begins actively pumping Ca^{++} back into its sacs.
2. As Ca^{++} is stripped from troponin molecules in the thin myofilaments, tropomyosin returns to its position, blocking actin's active sites.
3. Myosin cross bridges are prevented from binding to actin and thus can no longer sustain the contraction.
4. Since the thick and thin myofilaments are no longer connected, the muscle fiber may return to its longer, resting length.

TABLE 5-6   Muscles of facial expression and of mastication

Muscle	Origin	Insertion	Function	Nerve Supply
MUSCLES OF FACIAL EXPRESSION				
Epicranius (occipitofrontalis)	Occipital bone	Tissues of eyebrows	Raises eyebrows, wrinkles forehead horizontally	Cranial nerve VII
Corrugator supercilii	Frontal bone (superciliary ridge)	Skin of eyebrow	Wrinkles forehead vertically	Cranial nerve VII
Orbicularis oculi	Encircles eyelid		Closes eye	Cranial nerve VII
Zygomaticus major	Zygomatic bone	Angle of mouth	Laughing (elevates angle of mouth)	Cranial nerve VII
Orbicularis oris	Encircles mouth		Draws lips together	Cranial nerve VII
Buccinator	Maxillae	Skin of sides of mouth	Permits smiling	Cranial nerve VII
			Blowing, as in playing a trumpet	
MUSCLES OF MASTICATION				
Masseter	Zygomatic arch	Mandible (external surface)	Closes jaw	Cranial nerve V
Temporalis	Temporal bone	Mandible	Closes jaw	Cranial nerve V
Pterygoids (internal and external)	Undersurface of skull	Mandible (medial surface)	Grates teeth	Cranial nerve V

TABLE 5-7   Muscles that move the head

Muscle	Origin	Insertion	Function	Nerve Supply
Sternocleidomastoid	Sternum Clavicle	Temporal bone (mastoid process)	Flexes head (prayer muscle) One muscle alone, rotates head toward opposite side; spasm of this muscle alone or associated with trapezius called *torticollis* or *wryneck*	Accessory nerve
Semispinalis capitis	Vertebrae (transverse processes of upper six thoracic, articular processes of lower four cervical)	Occipital bone (between superior and inferior nuchal lines)	Extends head; bends it laterally	First five cervical nerves
Splenius capitis	Ligamentum nuchae Vertebrae (spinous processes of upper three or four thoracic)	Temporal bone (mastoid process) Occipital bone	Extends head Bends and rotates head toward same side as contracting muscle	Second, third, and fourth cervical nerves
Longissimus capitis	Vertebrae (transverse processes of upper six thoracic, articular processes of lower four cervical)	Temporal bone (mastoid process)	Extends head Bends and rotates head toward contracting side	Multiple innervation

TABLE 5-8   Muscles of the thorax

Muscle	Origin	Insertion	Function	Nerve Supply
External intercostals	Rib (lower border; forward fibers)	Rib (upper border of rib below origin)	Elevate ribs	Intercostal nerves
Internal intercostals	Rib (inner surface, lower border; backward fibers)	Rib (upper border of rib below origin)	Depress ribs	Intercostal nerves
Diaphragm	Lower circumference of thorax (of rib cage)	Central tendon of diaphragm	Enlarges thorax, causing inspiration	Phrenic nerves

TABLE 5-9   Muscles of the abdominal wall

Muscle	Origin	Insertion	Function	Nerve Supply
External oblique	Ribs (lower eight)	Ossa coxae (iliac crest and pubis by way of inguinal ligament) Linea alba by way of an aponeurosis	Compresses abdomen Important postural function of all abdominal muscles is to pull front of pelvis upward, thereby flattening lumbar curve of spine; when these muscles lose their tone, common figure faults of protruding abdomen and lordosis develop	Lower seven intercostal nerves and iliohypogastric nerves
Internal oblique	Ossa coxae (iliac crest and inguinal ligament) Lumbodorsal fascia	Ribs (lower three) Pubic bone Linea alba	Same as external oblique	Last three intercostal nerves; iliohypogastric and ilioinguinal nerves
Transversus abdominis	Ribs (lower six) Ossa coxae (iliac crest, inguinal ligament) Lumbodorsal fascia	Pubic bone Linea alba	Same as external oblique	Last five intercostal nerves; iliohypogastric and ilioinguinal nerves
Rectus abdominis	Ossa coxae (pubic bone and symphysis pubis)	Ribs (costal cartilage of fifth, sixth, and seventh ribs) Sternum (xiphoid process)	Same as external oblique; because abdominal muscles compress abdominal cavity, they aid in straining, defecation, forced expiration, childbirth, etc.; abdominal muscles are antagonists of diaphragm, relaxing as it contracts and vice versa Flexes trunk	Last six intercostal nerves

TABLE 5-10   Muscles of the pelvic floor

Muscle	Origin	Insertion	Function	Nerve Supply
Levator ani	Pubis and spine of ischium	Coccyx	Together with coccygeus muscles form floor of pelvic cavity and support pelvic organs	Pudendal nerve
Ischiocavernosus	Ischium	Penis or clitoris	Compress base of penis or clitoris	Perineal nerve
Bulbospongiosus				
Male	Bulb of penis	Perineum and bulb of penis	Constricts urethra and erects penis	Pudendal nerve
Female	Perineum	Base of clitoris	Erects clitoris	Pudendal nerve
Deep transverse perinei	Ischium	Central tendon (median raphe)	Support pelvic floor	Pudendal nerve
Sphincter urethrae	Pubic ramus	Central tendon (median raphe)	Constrict urethra	Pudendal nerve
Sphincter externus anii	Coccyx	Central tendon (median raphe)	Close anal canal	Pudendal and S4

TABLE 5-11   Muscles acting on the shoulder girdle

Muscle	Origin	Insertion	Function	Nerve Supply
Trapezius	Occipital bone (protuberance)	Clavicle	Raises or lowers shoulders and shrugs them	Spinal accessory; second, third, and fourth cervical nerves
	Vertebrae (cervical and thoracic)	Scapula (spine and acromion)	Extends head when occiput acts as insertion	
Pectoralis minor	Ribs (second to fifth)	Scapula (coracoid)	Pulls shoulder down and forward	Medial and lateral anterior thoracic nerves
Serratus anterior	Ribs (upper eight or nine)	Scapula (anterior surface, vertebral border)	Pulls shoulder forward; abducts and rotates it upward	Long thoracic nerve
Levator scapulae	C1-C4 (transverse processes)	Scapula (superior angle)	Elevates and retracts scapula and abducts neck	Dorsal scapular nerve
Rhomboideus				
Major	T1-T4	Scapula (medial border)	Retracts, rotates, fixes scapula	Dorsal scapular nerve
Minor	C6-C7	Scapula (medial border)	Retracts, rotates, elevates, and fixes scapula	Dorsal scapular nerve

TABLE 5-12   Muscles that move the upper arm

Muscle	Origin	Insertion	Function	Nerve Supply
Pectoralis major	Clavicle (medial half) Sternum Costal cartilages of true ribs	Humerus (greater tubercle)	Flexes upper arm Adducts upper arm anteriorly; draws it across chest	Medial and lateral anterior thoracic nerves
Latissimus dorsi	Vertebrae (spines of lower thoracic, lumbar, and sacral) Ilium (crest) Lumbodorsal fascia	Humerus (intertubercular groove)	Extends upper arm Adducts upper arm posteriorly	Thoracodorsal nerve
Deltoid	Clavicle Scapula (spine and acromion)	Humerus (lateral side about halfway down—deltoid tubercle)	Abducts upper arm Assists in flexion and extension of upper arm	Axillary nerve
Coracobrachialis	Scapula (coracoid process)	Humerus (middle third, medial surface)	Adduction; assists in flexion and medial rotation of arm	Musculocutaneous nerve
Supraspinatus	Scapula (supraspinous fossa)	Humerus (greater tubercle)	Assists in abducting arm	Suprascapular nerve
Teres minor	Scapula (axillary border)	Humerus (greater tubercle)	Rotates arm outward	Axillary nerve
Teres major	Scapula (lower part, axillary border)	Humerus (upper part, anterior surface)	Assists in extension, adduction; and medial rotation of arm	Lower subscapular nerve
Infraspinatus	Scapula (infraspinatus border)	Humerus (greater tubercle)	Rotates arm outward	Suprascapular nerve
Subscapularis	Scapula (subscapular fossa)	Humerus (lesser tubercle)	Medial rotation	Subscapular nerve

TABLE 5-13   Muscles that move the forearm

Muscle	Origin	Insertion	Function	Nerve Supply
Biceps brachii	Scapula (supraglenoid tuberosity)	Radius (tubercle at proximal end)	Flexes supinated forearm	Musculocutaneous nerve
	Scapula (coracoid)		Supinates forearm and hand	
Brachialis	Humerus (distal half, anterior surface)	Ulna (front of coronoid process)	Flexes pronated forearm	Musculocutaneous nerve
Brachioradialis	Humerus (above lateral epicondyle)	Radius (styloid process)	Flexes semipronated or semisupinated forearm; supinates forearm and hand	Radial nerve
Triceps brachii	Scapula (infraglenoid tuberosity)	Ulna (olecranon process)	Extends lower arm	Radial nerve
	Humerus (posterior surface—lateral head above radial groove; medial head, below)			
Pronator teres	Humerus (medial epicondyle)	Radius (middle third of lateral surface)	Pronates and flexes forearm	Median nerve
	Ulna (coronoid process)			
Pronator quadratus	Ulna (distal fourth, anterior surface)	Radius (distal fourth, anterior surface)	Pronates forearm	Median nerve
Supinator	Humerus (lateral epicondyle)	Radius (proximal third)	Supinates forearm	Radial nerve
	Ulna (proximal fifth)			

TABLE 5-14 Muscles that move the wrist, hand, and fingers

Muscle	Origin	Insertion	Function	Nerve Supply
Flexor carpi radialis	Humerus (medial epicondyle)	Second metacarpal (base of)	Flexes hand Flexes forearm	Median nerve
Palmaris longus	Humerus (medial epicondyle)	Fascia of palm	Flexes hand	Median nerve
Flexor carpi ulnaris	Humerus (medial epicondyle) Ulna (proximal two thirds)	Pisiform bone Third, fourth, and fifth metacarpals	Flexes hand Adducts hand	Ulnar nerve
Extensor carpi radialis longus	Humerus (ridge above lateral epicondyle)	Second metacarpal (base of)	Extends hand Abducts hand (moves toward thumb side when hand supinated)	Radial nerve
Extensor carpi radialis brevis	Humerus (lateral epicondyle)	Second, third metacarpals (bases of)	Extends hand	Radial nerve
Extensor carpi ulnaris	Humerus (lateral epicondyle) Ulna (proximal three fourths)	Fifth metacarpal (base of)	Extends hand Adducts hand (moves toward little finger side when hand supinated)	Radial nerve
Flexor digitorum profundus	Ulna (anterior surface)	Distal phalanges (fingers 2 to 5)	Flexes distal interphalangeal joints	Median and ulnar nerves
Flexor digitorum superficialis	Humerus (medial epicondyle) Radius Ulna (coronoid process)	Tendons of fingers	Flexes fingers	Median nerve
Extensor digitorum	Humerus (lateral epicondyle)	Phalanges (fingers 2 to 5)	Extends fingers	Radial nerve
Opponens pollicis	Greater multangular	Thumb metacarpal	Opposes thumb to fingers	Median nerve

TABLE 5-15   Muscles that move the thigh

Muscle	Origin	Insertion	Function	Nerve Supply
Iliopsoas (iliacus, psoas major, and psoas minor)	Ilium (iliac fossa) Vertebrae (bodies of twelfth thoracic to fifth lumbar)		Flexes thigh Flexes trunk (when femur acts as origin)	Femoral and second to fourth lumbar nerves
Rectus femoris	Ilium (anterior, inferior spine)	Tibia (by way of patellar tendon)	Flexes thigh Extends lower leg	Femoral nerve
Gluteal group				
Maximus	Ilium (crest and posterior surface) Sacrum and coccyx (posterior surface) Sacrotuberous ligament	Femur (gluteal tuberosity) Iliotibial tract	Extends thigh—rotates outward	Inferior gluteal nerve
Medius	Ilium (lateral surface)	Femur (greater trochanter)	Abducts thigh—rotates outward; stabilizes pelvis on femur	Superior gluteal nerve
Minimus	Ilium (lateral surface)	Femur (greater trochanter)	Abducts thigh; stabilizes pelvis on femur Rotates thigh medially	Superior gluteal nerve
Tensor fasciae latae	Ilium (anterior part of crest)	Tibia (by way of iliotibial tract)	Abducts thigh Tightens iliotibial tract	Superior gluteal nerve
Adductor group				
Brevis	Pubic bone	Femur (linea aspera)	Adducts thigh	Obturator nerve
Longus	Pubic bone	Femur (linea aspera)	Adducts thigh	Obturator nerve
Magnus	Pubic bone	Femur (linea aspera)	Adducts thigh	Obturator nerve
Gracilis	Pubic bone (just below symphysis)	Tibia (medial surface behind sartorius)	Adducts thigh and flexes and adducts leg	Obturator nerve

TABLE 5-16   Muscles that move the lower leg

Muscle	Origin	Insertion	Function	Nerve Supply
Quadriceps femoris group				
Rectus femoris	Ilium (anterior inferior spine)	Tibia (by way of patellar tendon)	Flexes thigh Extends leg	Femoral nerve
Vastus lateralis	Femur (linea aspera)	Tibia (by way of patellar tendon)	Extends leg	Femoral nerve
Vastus medialis	Femur	Tibia (by way of patellar tendon)	Extends leg	Femoral nerve
Vastus intermedius	Femur (anterior surface)	Tibia (by way of patellar tendon)	Extends leg	Femoral nerve
Sartorius	Coxal (anterior, superior iliac spines)	Tibia (medial surface of upper end of shaft)	Adducts and flexes leg Permits crossing of legs tailor fashion	Femoral nerve
Hamstring group				
Biceps femoris	Ischium (tuberosity)	Fibula (head of)	Flexes leg	Hamstring nerve (branch of sciatic nerve)
	Femur (linea aspera)	Tibia (lateral condyle)	Extends thigh	Hamstring nerve
Semitendinosus	Ischium (tuberosity)	Tibia (proximal end, medial surface)	Extends thigh	Hamstring nerve
Semimembranosus	Ischium (tuberosity)	Tibia (medial condyle)	Extends thigh	Hamstring nerve

TABLE 5-17   Muscles that move the foot

Muscle	Origin	Insertion	Function	Nerve Supply
Tibialis anterior	Tibia (lateral condyle of upper body)	Tarsal (first cuneiform) Metatarsal (base of first)	Flexes foot Inverts foot	Common and deep peroneal nerves
Gastrocnemius	Femur (condyles)	Tarsal (calcaneus by way of Achilles tendon)	Extends foot Flexes lower leg	Tibial nerve (branch of sciatic nerve)
Soleus	Tibia (underneath gastrocnemius) Fibula	Tarsal (calcaneus by way of Achilles tendon)	Extends foot (plantar flexion)	Tibial nerve
Peroneus longus	Tibia (lateral condyle) Fibula (head and shaft)	First cuneiform Base of first metatarsal	Extends foot (plantar flexion) Everts foot	Common peroneal nerve
Peroneus brevis	Fibula (lower two thirds of lateral surface of shaft)	Fifth metatarsal (tubercle, dorsal surface)	Everts foot Flexes foot	Superficial peroneal nerve
Peroneus tertius	Fibula (distal third)	Fourth and fifth metatarsals (bases of)	Flexes foot Everts foot	Deep peroneal nerve
Extensor digitorum longus	Tibia (lateral condyle) Fibula (anterior surface)	Second and third phalanges (four lateral toes)	Dorsiflexion of foot; extension of toes	Deep peroneal nerve

TABLE 5-18 Spinal nerves and peripheral branches

Spinal Nerves	Plexuses Formed from Anterior Rami	Spinal Nerve Branches from Plexuses	Parts Supplied
CERVICAL 1 2 3 4	Cervical plexus	Lesser occipital Greater auricular Cutaneous nerve of neck Supraclavicular nerves Branches to muscles	Sensory to back of head, front of neck, and upper part of shoulder; motor to numerous neck muscles
		Phrenic nerve	Diaphragm
CERVICAL 5 6 7 8 **THORACIC (OR DORSAL)** 1	Brachial plexus	Suprascapular and dorsoscapular	Superficial muscles* of scapula
		Thoracic nerves, medial and lateral branches	Pectoralis major and minor
		Long thoracic nerve	Serratus anterior
		Thoracodorsal	Latissimus dorsi
		Subscapular	Subscapular and teres major muscles
		Axillary (circumflex)	Deltoid and teres minor muscles and skin over deltoid
2 3 4 5 6 7 8 9 10 11 12	No plexus formed; branches run directly to intercostal muscles and skin of thorax	Musculocutaneous	Muscles of front of arm (biceps brachii, coracobrachialis, and brachialis) and skin on outer side of forearm
		Ulnar	Flexor carpi ulnaris and part of flexor digitorum profundus; some of muscles of hand; sensory to medial side of hand, little finger, and medial half of fourth finger
		Median	Rest of muscles of front of forearm and hand; sensory to skin of palmar surface of thumb, index, and middle fingers
		Radial	Triceps muscle and muscles of back of forearm; sensory to skin of back of forearm and hand
		Medial cutaneous	Sensory to inner surface of arm and forearm
		Iliohypogastric *Sometimes fused*	Sensory to anterior abdominal wall
		Ilioinguinal	Sensory to anterior abdominal wall and external genitalia; motor to muscles of abdominal wall
		Genitofemoral	Sensory to skin of external genitalia and inguinal region
LUMBAR 1 2 3 4 5 **SACRAL** 1 2 3 4 5 **COCCYGEAL** 1	Lumbosacral plexus	Lateral femoral cutaneous	Sensory to outer side of thigh
		Femoral	Motor to quadriceps, sartorius, and iliacus muscles; sensory to front of thigh and medial side of lower leg (saphenous nerve)
		Obturator	Motor to adductor muscles of thigh
		Tibial† (medial popliteal)	Motor to muscles of calf of leg; sensory to skin of calf of leg and sole of foot
		Common peroneal (lateral popliteal)	Motor to evertors and dorsiflexors of foot; sensory to lateral surface of leg and dorsal surface of foot
		Nerves to hamstring muscles	Motor to muscles of back of thigh
		Gluteal nerves	Motor to buttock muscles and tensor fasciae latae
		Posterior femoral cutaneous	Sensory to skin of buttocks, posterior surface of thigh, and leg
	Coccygeal plexus	Pudendal nerve	Motor to perineal muscles; sensory to skin of perineum

*Although nerves to muscles are considered motor, they do contain some sensory fibers that transmit proprioceptive impulses.

†Sensory fibers from the tibial and peroneal nerves unite to form the *medial cutaneous* (or sural) *nerve* that supplies the calf of the leg and the lateral surface of the foot. In the thigh the tibial and common peroneal nerves are usually enclosed in a single sheath to form the *sciatic nerve,* the largest nerve in the body with a width of approximately ¾ of an inch. About two thirds of the way down the posterior part of the thigh, it divides into its component parts. Branches of the sciatic nerve extend into the hamstring muscles.

TABLE 5-19 Hormones of the pituitary gland (hypophysis)

Hormone	Source	Target	Principal action
Growth hormone (GH) [somatotropin (STH)]	Adenohypophysis (acidophils)	General	Promotes growth by stimulating protein anabolism and fat mobilization
Prolactin (PRL) [lactogenic hormone]	Adenohypophysis (acidophils)	Mammary glands (alveolar secretory cells)	Promotes milk secretion
Thyroid-stimulating hormone (TSH)*	Adenohypophysis (basophils)	Thyroid gland	Stimulates development and secretion in the thyroid gland
Adrenocorticotropic hormone (ACTH)*	Adenohypophysis (basophils)	Adrenal cortex	Promotes development and secretion in the adrenal cortex
Follicle-stimulating hormone (FSH)*	Adenohypophysis (basophils)	Gonads (primary sex organs)	Female: promotes development of ovarian follicle; stimulates estrogen secretion Male: promotes development of testis; stimulates sperm production
Luteinizing hormone (LH)*	Adenohypophysis (basophils)	Gonads and mammary glands	Female: triggers ovulation; promotes development of corpus luteum Male: stimulates production of testosterone
Melanocyte-stimulating hormone (MSH)	Adenohypophysis (basophils)	Skin (melanocytes); adrenal glands	Exact function uncertain; may stimulate production of melanin pigment in skin; may maintain adrenal sensitivity
Antidiuretic hormone (ADH)	Neurohypophysis	Kidney	Promotes water retention by kidney tubules
Oxytocin (OT)	Neurohypophysis	Uterus and mammary glands	Stimulates uterine contractions; stimulates ejection of milk into mammary ducts

*Tropic hormones

TABLE 5-20   Classes of blood cells

Cell Type	Description	Function	Life Span
Erythrocyte	7 μm in diameter; concave disk shape; entire cell stains pale pink; no nucleus	Transportation of respiratory gases (O_2 and CO_2)	105 to 120 days
Neutrophil	12-15 μm in diameter; spherical shape; multilobed nucleus; small, pink-purple staining cytoplasmic granules	Cellular defense—phagocytosis of small pathogenic microorganisms	Hours to 3 days
Basophil	11-14 μm in diameter; spherical shape; generally two lobed nucleus; large purple staining cytoplasmic granules	Secretes heparin (anticoagulant) and histamine (important in inflammatory response)	Hours to 3 days
Eosinophil	10-12 μm in diameter; spherical shape; generally two-lobed nucleus; large orange-red staining cytoplasmic granules	Cellular defense—phagocytosis of large pathogenic microorganisms such as protozoa and parasitic worms; releases antiinflamatory substances in allergic reactions	10 to 12 days
Lymphocyte	6-9 μm in diameter; spherical shape; round (single lobe) nucleus; small lymphocytes have scant cytoplasm	Humoral defense—secretes antibodies; involved in immune system response and regulation	Days to years
Monocyte	12–17 μm in diameter; spherical shape; nucleus generally kidney-bean or "horse-shoe" shaped with convoluted surface; ample cytoplasm often "steel blue" in color	Capable of migrating out of the blood to enter tissue spaces as a *macrophage*—an aggressive phagocytic cell capable of ingesting bacteria, cellular debris, and cancerous cells	Months
Platelet	2-5 μm in diameter; irregularly shaped fragments; cytoplasm contains very small pink staining granules	Releases clot activating substances and helps in formation of actual blood clot by forming platelet "plugs"	7 to 10 days

INDEX